30-SECOND
ANATOMY

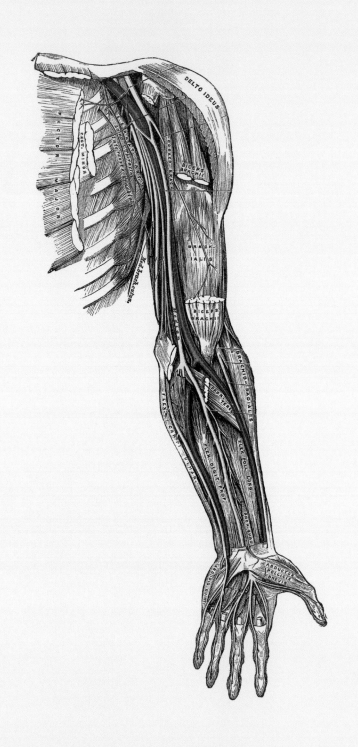

30-SECOND ANATOMY

The 50 most important structures and systems in the human body, each explained in half a minute

Editor
Gabrielle M. Finn

Contributors
Judith Barbaro-Brown
Jo Bishop
Andrew Chaytor
Gabrielle M. Finn
December S. K. Ikah
Marina Sawdon
Claire France Smith

METRO BOOKS
NEW YORK

METRO BOOKS
New York

An Imprint of Sterling Publishing
387 Park Avenue South
New York, NY 10016

METRO BOOKS and the distinctive Metro Books logo are trademarks of Sterling Publishing Co., Inc.

This book was conceived, designed, and produced by

Ivy Press
210 High Street, Lewes,
East Sussex BN7 2NS, U.K.
www.ivy-group.co.uk

Creative Director **Peter Bridgewater**
Publisher **Jason Hook**
Editorial Director **Caroline Earle**
Art Director **Michael Whitehead**
Designer **Ginny Zeal**
Illustrator **Ivan Hissey**
Profiles Text **Viv Croot**
Glossaries Text **Charles Phillips**
Project Editor **Jamie Pumfrey**

ISBN: 978-1-4351-4065-3

For information about custom editions, special sales, and premium and corporate purchases, please contact Sterling Special Sales at 800-805-5489 or specialsales@sterlingpublishing.com.

Manufactured in China

Color origination by Ivy Press Reprographics

4 6 8 10 9 7 5 3

www.sterlingpublishing.com

CONTENTS

INTRODUCTION
Gabrielle M. Finn

Anatomy is both within and all around us. By learning a little anatomy, we come to understand how our bodies are built: An anatomical drawing that depicts the bones, muscles, ligaments, tendons, and organs of the body is a map of the inner landscape we all share. Yet at the same time, our experience of the body, our knowledge of its skeleton and organs, informs the way we see the world. For this reason, human anatomy has a widespread symbolism in popular culture, from the hearts printed on Valentine's Day cards to the skull as a symbol of danger.

Gruesome history

Traditionally, many people may have seen anatomy as an academic discipline, of interest to only medical students, but in recent years the subject has had a boom in popularity. This is due largely to its entry into the public arena through touring exhibitions of cadaveric specimens and televised human dissections by anatomists, such as Gunther von Hagens and Alice Roberts. Behind this new interest lies a long and gruesome history.

The origins of anatomical study were in animal vivisection and the dissection of human corpses. The ancient Greek physician Galen based his ideas of human anatomy on knowledge gained from dissections and vivisections of pigs and primates. The Italian Renaissance artist Leonardo da Vinci, creator of *The Last Supper* and the *Mona Lisa*, was famed for his anatomical drawings and derived his knowledge of the inside of the human body from working with corpses supplied by doctors in hospitals in Milan and Florence. Anatomy has also been associated with crime, as in the case of the 19th-century murders perpetrated in Scotland by William Burke and William Hare. A pair of Irish immigrants, they robbed graves and embarked on a serial-killing spree in 1827–1828 in order to sell corpses to Dr. Robert Knox, an anatomy lecturer with students from Edinburgh University medical school. The pair were caught: Hare was granted immunity for testifying, and Burke was hanged on January 28, 1829; ironically, his remains ended up in the medical school's anatomy museum.

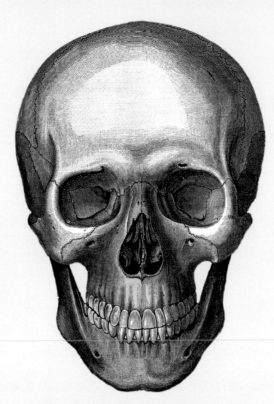

Anatomical blueprint
Despite the fact that anatomy describes the blueprint for the living body, one of the subject's most common associations is death—the bones that structure our living bodies are our final physical remains.

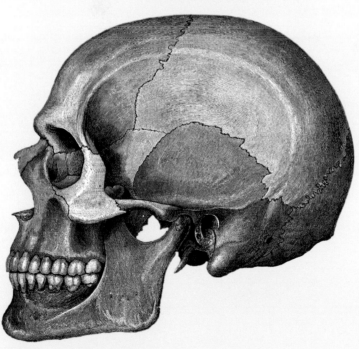

Evolution and variation

Anatomy is an ancient discipline, and you might think that there is nothing new to know in the field. Yet, remarkably, human anatomy continues to evolve. This evolution is very slow, but it exists and persists. Take, for example, the coccygeal bones at the base of the spine: These used to be the point at which the human tail started. Another example of continual evolution is the palmaris longus muscle in the forearm; due to its limited function, this muscle has become redundant, and evolution in some individuals has resulted in its absence in around 15 percent of people.

One of the biggest challenges facing anyone who wants to study anatomy is anatomical variation. As we have seen, anatomy provides a blueprint for how all bodies are structured; however, in a world with a population heading for an estimated 7 billion people, variation will and does exist. A primary example is found in the arteries within the pelvis; there are 54 known variations of how these vessels distribute themselves. Moreover, variation exists not only from person to person, but also from side to side within an individual body. Some people have a larger ear on the right than on the left; or a person may have a single horseshoe-shaped kidney instead of the normal two kidneys; or the pathways followed by nerves may vary from the accepted convention. This book presents the most commonly encountered anatomy.

Anatomy—systems and functions

Anatomy has its own technical language, in which muscles and bones have lengthy Latin or Greek names. Simple physical actions, such as the movement of the lower limb (leg), have multiple anatomical descriptions depending on the direction of the movement. There are in excess of 200 bones, 600 muscles, and numerous veins, arteries, and nerves. Don't let that put you off. This book does not attempt to explain the location and function of each individual structure; instead, it breaks the body down into functional systems and describes the 50 most relevant components, using illustrations and avoiding complex terminology.

Another consideration is that anatomical structures—whether a single muscle in the thigh or a digestive organ, such as the stomach—do not work alone. Although the text maps out individualized functions for each structure, the reality is that everything works together. The function of one organ might rely on a hormone produced by another, or the movement at a joint may result from the actions of three or four muscles working together. Think about the bigger picture.

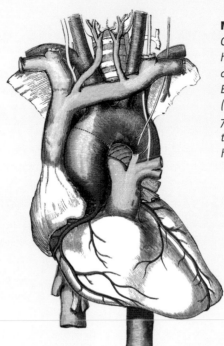

Medical pioneers
Great breakthroughs in anatomy have often taken the form of correcting earlier errors. The Englishman William Harvey (1578–1625)—featured on pages 74–75—was the first to establish the true role in the body of the heart (top) and lungs (below).

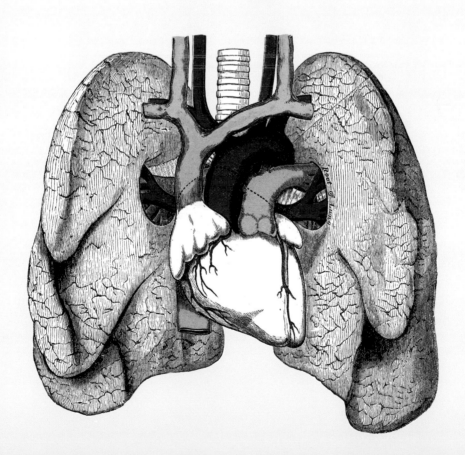

How this book works

Traditionally, anatomy is regarded as the study of the body's form or structure, whereas physiology describes its functioning. However, form and function cannot be mutually exclusive. This book does not divorce the two; it describes both how the body is built and how it works.

The construction of the human body can be described in one of two ways: regionally or systematically. A regional approach to writing about anatomy would be to describe the individual areas of the body, such as the leg, the arm, or the head; a systematic approach would describe the body systems—digestive, musculoskeletal, reproductive, and so on. The approach in *30-Second Anatomy* is based on these systems. The book is organized into seven chapters, each addressing a system of the body. By the end, it is hoped that you'll have dissected your way through the body, learning the bare bones of its anatomy, in a simple, easy-to-follow manner.

Each anatomical component is presented as a 30-second anatomy. Accompanying this is the 3-second incision for those who simply want a quick slice. The 3-minute dissection that follows serves to illustrate the claim of Sigmund Freud, the Austrian neurologist and founder of psychoanalysis, that "Anatomy is destiny." It provides examples of weird and wonderful aspects of human anatomy and describes what happens when bodily structures go wrong.

The first chapter deals with the skeletal system. The human body is constructed around the skeleton; the bones are a scaffolding onto which everything else is built. The second chapter looks at the muscular system and how humans move. The next two chapters are oriented around the main organs of the cardiovascular and digestive systems, addressing key functions, such as breathing and eating, and how the blood is pumped around the body. Next comes a tour of the special senses—skin, sight, and hearing, to name but a few. The sixth chapter considers overall control of bodily function—the brain and the nervous system. Finally, we end with a nod to the circle of life by examining the reproductive system. Within each chapter you will also find a profile of a key anatomist.

The structure of this book is such that you can dip in and out, reading the odd entry here and there, or you can go through it system by system, or read it cover to cover. So why not lift the hood on your anatomy—and read on to enjoy finding out the mechanics of how you work!

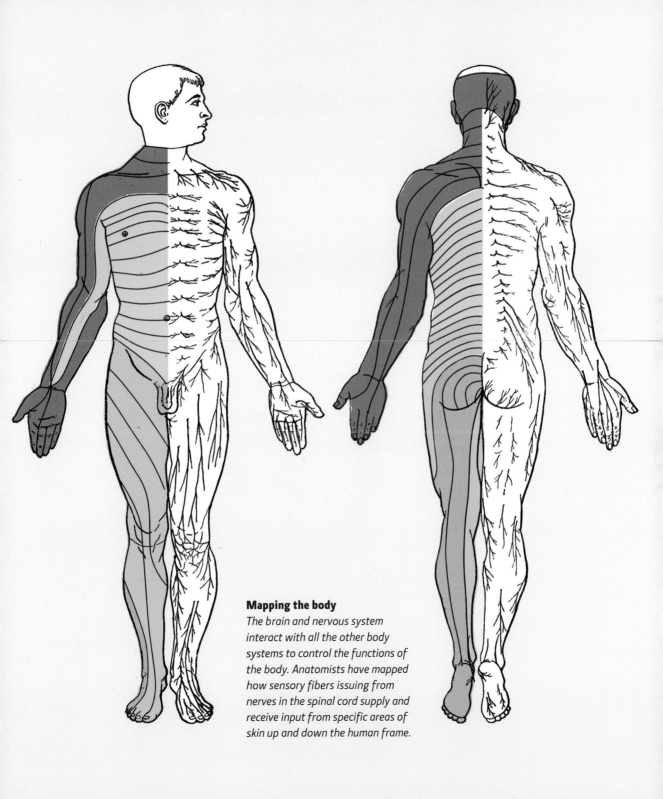

Mapping the body

The brain and nervous system interact with all the other body systems to control the functions of the body. Anatomists have mapped how sensory fibers issuing from nerves in the spinal cord supply and receive input from specific areas of skin up and down the human frame.

THE SKELETAL SYSTEM

THE SKELETAL SYSTEM
GLOSSARY

cartilage Type of connective tissue consisting largely of water, minerals, and the proteins collagen and elastin. There are three types: hyaline cartilage covers the surface of joints and allows bones to move easily against one another; elastic cartilage provides structure for body parts that need to be pliable, such as the ears or nose; and fibrocartilage provides solidity in the spine's intervertebral disks or the knee's menisci.

cartilaginous joint One of three types of joints in the body (with fibrous and synovial joints); cartilaginous joints, held together by flexible cartilage that allows for limited movement, are found in between the vertebrae in the spine.

cortical bone Dense, strong outer layer surrounding a bone's relatively light honeycomb-like inner structure. Also known as compact bone, it is called cortical because it forms the cortex (outer coating) of the bone. Cortical bone makes up 80 percent of the weight of the human skeleton.

femur Also known as the thighbone, bone that reaches from the hip joint to the knee joint; typically 19 inches (48 cm) in length and capable of supporting up to 30 times a person's weight, the femur is the longest and strongest in the body.

fibrous joint Found only in the skull, a type of joint that connects bones with fibrous tissue and allows for no movement.

flat bone Bone that takes the form of a wide plate to provide protection for a body organ or a surface for the attachment of muscles; examples include the sternum (breastbone) and scapula (shoulder blade). Flat bone is one of five bone types: long, short, flat, irregular, and sesamoid.

humerus Long bone that connects the scapula (shoulder blade) to the radius and ulna of the forearm.

irregular bone Bone with unusual form that cannot be classified within the categories of long, short, flat, and sesamoid bones. Examples include the vertebrae, which protect the spinal cord.

ligament Type of connective tissue that connects bones to one another and limits movement between them.

limb Appendage on the side of the body. Humans have four limbs—two upper (arms) and two lower (legs). Each has four sections: shoulder, arm, forearm, and hand in the upper limb; buttock, thigh, leg, and foot in the lower limb.

long bone Elongated bone, one longer than it is wide. Examples include the humerus and radius in the arm, and the femur and tibia in the leg. Small bones, such as the phalanges (in the fingers and toes), are classified as long bones because of their elongated shape.

sesamoid bone Rounded piece of bone, usually set within tendon, often less than $1/4$ inch (5 mm) in length. A larger example of a sesamoid bone is the kneecap (patella), which is embedded in the tendon of the thigh's extensor muscle and serves to protect the knee joint.

short bone Bone as wide as it is long; examples include the carpals and tarsals in the wrist and foot.

synovial joint Type of joint which is designed to facilitate movement; it is filled with a lubricating fluid. There are six types of synovial joint: pivot joints allow rotation, as, for example, in the neck; hinge joints allow for a body part to be straightened or retracted, as in the elbow; ball-and-socket joints, for example, in the hip, allow for radial movement; saddle joints, found in the thumb, permit up-and-down and forward-and-backward movement, but not rotation; plane joints, such as those between the intertarsal bones of the feet, allow for gliding of bones; and ellipsoid joints, for instance, in the wrist, allow for the same movement as ball-and-socket joints, but to a lesser extent.

tendon Band of connective tissue that attaches a muscle to a bone.

trabecular bone Lighter inner part of bone, also called cancellous bone, which is protected by a stronger outer layer (cortical bone); trabecular bone often contains red bone marrow, in which red blood cells are produced.

vertebrae One of the interconnecting bones that form the spinal column. Children have 33 vertebrae, but in adults 5 unite to make the sacrum and 4 combine to form the coccyx, reducing the number of bones to 26.

TYPES OF
BONE TISSUE

the 30-second anatomy

Bones combine to form the
skeleton, the supporting framework on
which everything in the body rests—without
their bones, humans would have difficulty
standing upright. Bone is a hard material made
by specialized bone cells called osteoclasts,
which combine minerals, such as calcium, with
phosphate and a protein called collagen. The
osteoclasts can form two types of bone:
trabecular bone, which looks spongy, is relatively
light and makes up the inside of most bones;
while cortical bone forms a very dense and
strong coat around the trabecular bone. The
combination of trabecular and cortical bone
makes the skeleton strong and keeps it light.
If this were not the case, humans would need
much bigger muscles to move the skeleton
around, as well as a good deal more food to
provide the necessary extra energy. Trabecular
bone has another important function: It acts
as a reservoir for calcium, which can be
extracted from the bone and used elsewhere
when other body systems are running low.
The bones are constantly changing and
renewing themselves—adult humans completely
replace their skeleton every ten years or so.

RELATED TOPICS
See also
THE SKULL
page 22
THE SPINE & RIB CAGE
page 24
THE PELVIS
page 28
THE LOWER LIMBS
page 30
THE UPPER LIMBS
page 32

3-SECOND BIOGRAPHY
HEROPHILOS
ca. 335–280 BCE
A Greek physician, the first
to systematically perform
scientific dissections of human
corpses, who is deemed to be
the first anatomist

30-SECOND TEXT
Judith Barbaro-Brown

3-SECOND INCISION
Bone is the scaffolding
for the human body, and
there are two types: hard
cortical bone and light
trabecular bone.

3-MINUTE DISSECTION
The skeleton consists of
206 bones and makes up
around 40 percent of body
weight. Bones come in all
shapes and sizes, from the
long bones in the limbs,
to the flat and sesamoid
bones in the spine, skull,
hands, and feet. The femur
in the lower limb is the
largest at around 19 inches
(48 cm) long. The smallest
is the stapes, around
$\frac{1}{10}$ inch (2.5 mm) long and
weighing around 4 mg.

*A "broken leg"—for a
skier or anyone else—
may range from a crack
in the outer cortical
layer to a complete
fracture of the bone
into two pieces.*

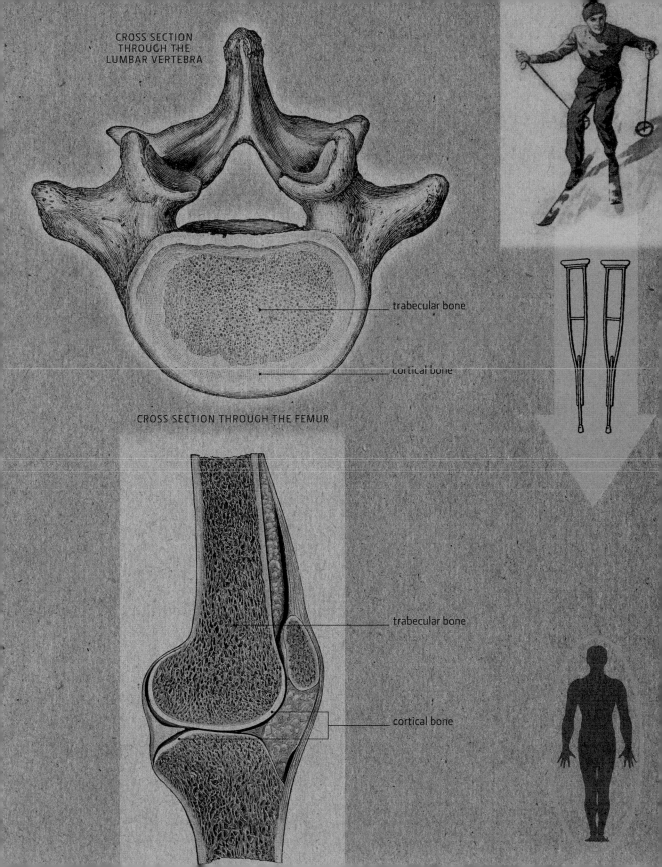

CROSS SECTION
THROUGH THE
LUMBAR VERTEBRA

trabecular bone

cortical bone

CROSS SECTION THROUGH THE FEMUR

trabecular bone

cortical bone

THE BONE JOINTS

the 30-second anatomy

Classified according to how the bones involved are held together, there are three classes of joints: fibrous, cartilaginous, and synovial. In fibrous joints, two bones are united by fibrous connective tissue with no space between them; these joints allow for minimal movement. Skull bones are joined like this. Cartilaginous joints are held together by flexible cartilage. Some of these joints are temporary, and eventually they completely fuse when growth is complete; however, in other places, such as at the front of the pelvis, the joint is permanent—these joints allow for a small amount of movement. In women, the cartilaginous pelvic joint enables the pelvis to become wider during pregnancy. Synovial joints allow for much more movement, and they are filled with a lubricating fluid that can also be a shock absorber. The fluid comes from a synovial membrane that encases the whole joint; around this is a tough joint capsule. The ends of the bones are covered in a smooth type of cartilage, which makes movement easy. Some joints contain specialized structures that provide added strength and stability—the knee has two of these (menisci)—and these are easily damaged when playing particularly physical sports, such as squash.

3-SECOND INCISION
Joints allow for the skeleton to move while also providing support and strength.

3-MINUTE DISSECTION
As people get older, their joints may wear out. The smooth surface of the bones becomes damaged and rough, making movement painful—a disorder known as osteoarthritis. This tends to occur in the joints that support body weight, such as the hips and knees, but can also be seen in the joints that are responsible for a lot of movement, such as fingers, spine, shoulders, and the neck.

RELATED TOPICS
See also
TYPES OF BONE TISSUE
page 16
THE LIGAMENTS, CARTILAGE & TENDONS
page 20
THE PELVIS
page 28
MOVEMENTS
page 42

3-SECOND BIOGRAPHY
RUFUS OF EPHESUS
ca. late 1st century CE
A Greek physician and the author of several medical treatises including the oldest surviving book on anatomical nomenclature

30-SECOND TEXT
Judith Barbaro-Brown

People do not need much movement at joints in the skull or pelvis, but make the most of flexible synovial joints in—for example—the neck.

FIBROUS JOINT

fibrous joints in skull bones

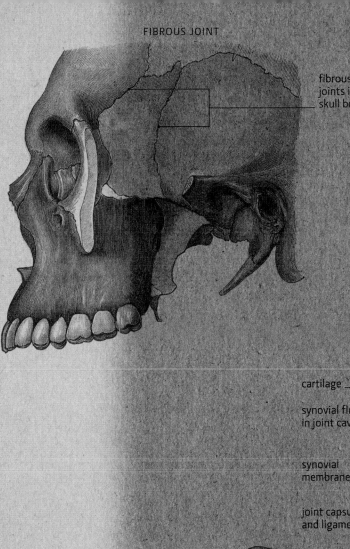

SYNOVIAL JOINT

cartilage

synovial fluid in joint cavity

synovial membrane

joint capsule and ligament

CARTILAGINOUS JOINT

cartilaginous joint in pubic bone

ligaments

THE LIGAMENTS, CARTILAGE & TENDONS

the 30-second anatomy

3-SECOND INCISION
Ligaments and cartilage hold the skeleton together, while tendons attach muscles to bones.

3-MINUTE DISSECTION
Ligaments, cartilage, and tendons all contain very high levels of the structural protein collagen, which can exist in a number of forms. In cartilage, collagen is very hard and can withstand high stress and forces. In ligaments, it is softer and allows for flexibility and stretching, while in tendons it is extremely strong and enables muscles to move very heavy bones.

Ligaments are made of collagen and fibrous tissue. They connect bones to one another and limit movement between the bones to increase stability. Where they cross at synovial joints, they contain proprioceptors, bundles of cells that detect the amount of movement in a joint. If there is a danger that the joint might become damaged, the proprioceptors dispatch a signal to the brain, which then sends instructions to the muscles to limit movement and so protect the joint. Cartilage is not as hard as bone and not as flexible as ligaments and muscles; it has no blood supply, which means that when damaged it is difficult to repair. Cartilage is usually associated with joints, where it provides a smooth surface so that bones can move easily against one another; however, it can also be found in the nose, ear, and respiratory system, where its job is to act as scaffolding. Tendons are tough bands of connective tissue that attach muscles to bones. Made of very strong collagen, they cannot contract or stretch, and they run from the muscle across a joint, attaching to a bone some distance from the muscle. When the muscle contracts it pulls on the tendon, which creates movement in the joint.

RELATED TOPICS
See also
THE BONE JOINTS
page 18
TYPES OF MUSCLE TISSUE
page 40
MOVEMENTS
page 42

30-SECOND TEXT
Judith Barbaro-Brown

Runners frequently strain the Achilles tendon in the lower leg. It is named after the ancient Greek hero Achilles, who was vulnerable in the heel.

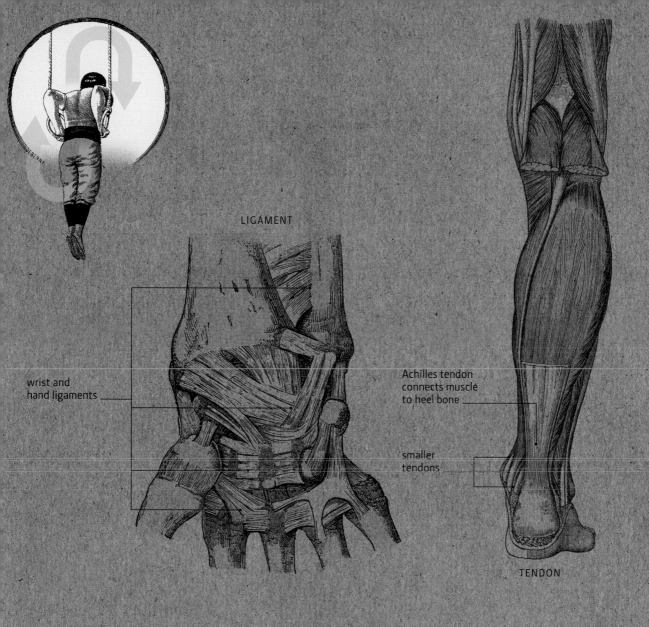

LIGAMENT

wrist and
hand ligaments

Achilles tendon
connects muscle
to heel bone

smaller
tendons

TENDON

CARTILAGE

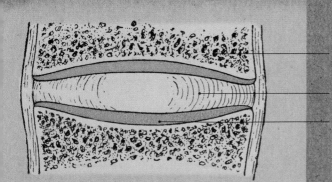

vertebra

fibrocartilage of intervertebral
disk provides solidity

cartilaginous end
plate covers joints

THE SKULL

the 30-second anatomy

People often assume that the skull, also known as the cranium, is one large bone. In fact, it consists of 22 bones. The skull can be divided into two parts: the cranial bones that protect the brain and the facial bones that form the face. The cranium is like a box and the brain nestles inside it. Within the base of the cranium are holes known as foramina, through which nerves and blood vessels transmit between the brain and the body. The largest is the foramen magnum, where the spinal cord exits the cranium. The bones of the adult skull are fused together by immovable fibrous joints, known as sutures. However, the bones of a baby's skull are not fused; this lets the bones overlap during delivery of the baby's head through the birth canal and permits postnatal growth. The skull's facial bones form cavities that house the organs relating to the senses—eyes, ears, nose, and mouth. The bones of the skull also have associated features, such as areas of extra bone (processes), that holds muscles and ligaments.

3-SECOND INCISION
The skull sits at the top of the spine, serving to house the brain and protect it from damage.

3-MINUTE DISSECTION
The facial bones form the solid framework onto which the soft tissue of the face is built. The shape of these bones determines a person's facial characteristics. Forensic scientists and artists can construct digital images of the superficial appearance of someone's face by measuring the bones of the skull. The measurements produce a map that can be used to create an impression of what the person might look like.

RELATED TOPICS
See also
THE FACIAL MUSCLES
page 44

3-SECOND BIOGRAPHY
WALTER J. FREEMAN
1895–1972
A U.S. physician, the first to perform a frontal lobotomy

30-SECOND TEXT
Gabrielle M. Finn

The frontal bone of the skull has a vertical part that corresponds with the forehead and a horizontal section that links to the top of the eye and nose cavities.

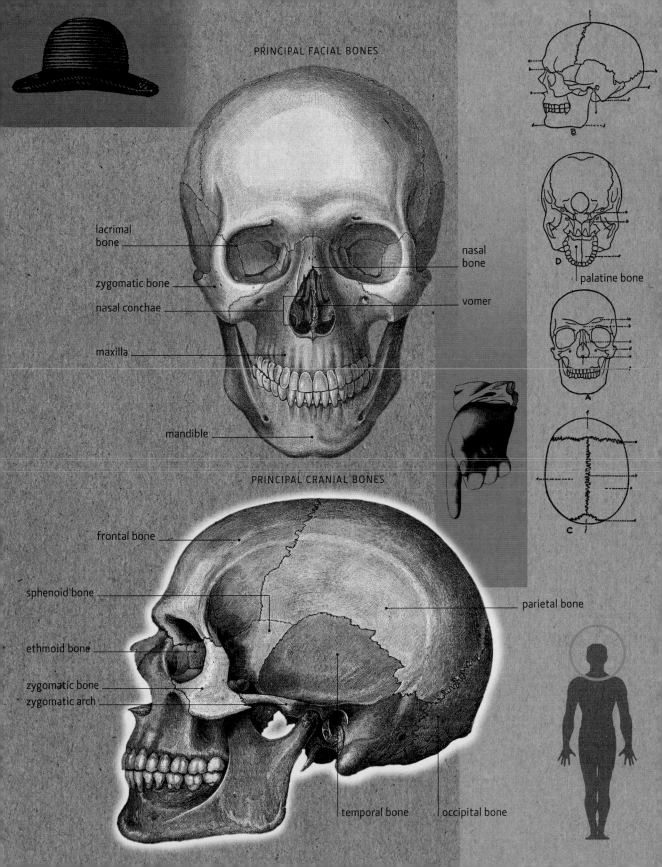

PRINCIPAL FACIAL BONES

lacrimal bone

nasal bone

zygomatic bone

nasal conchae

vomer

maxilla

mandible

palatine bone

PRINCIPAL CRANIAL BONES

frontal bone

sphenoid bone

parietal bone

ethmoid bone

zygomatic bone

zygomatic arch

temporal bone

occipital bone

THE SPINE
& RIB CAGE

the 30-second anatomy

3-SECOND INCISION
The spine supports the upper body, providing attachment for large muscles; the bones of the ribs form a cage that protects heart and lungs.

3-MINUTE DISSECTION
In some parts of the body, the spine is very flexible; in the neck, it provides a wide range of head movements, and in the lumbar area, it allows for mobility around the pelvis and lower back. Around the rib cage, however, the spine is relatively inflexible, because here its most important role is to provide anchorage for the ribs and protection for the heart and lungs.

The spine consists of 24 vertebral bones, separated by cushionlike intervertebral disks, and supported at the base by the sacrum. The shape of vertebrae varies depending on their position in the spine. There are 7 small, narrow cervical vertebrae, supporting the neck and skull; 12 thoracic vertebrae, all with very prominent spinous processes (projections from the spine) at the back, provide anchor points for the rib cage; 5 lumbar vertebrae—the thickest and biggest vertebrae—support the lower back and upper body. The sacrum is made from five fused vertebrae, and ends in the coccyx, which also consists of tiny fused vertebrae. Vertebrae are attached to each other by strong ligaments. Intervertebral disks are made from a thick, jellylike substance that provides shock absorption and flexibility. There are 12 pairs of ribs: the first 7 pairs, known as "true ribs," attach directly to the sternum via the costal cartilages; pairs 8 to 10, known as "false ribs," join with the costal cartilages of the ribs above; and pairs 11 and 12, the "floating ribs," are attached to only the spine and do not extend to the front of the body.

RELATED TOPICS
See also
THE LIGAMENTS, CARTILAGE & TENDONS
page 20
TYPES OF MUSCLE TISSUE
page 40
THE ABDOMINAL & BACK MUSCLES
page 54
THE RESPIRATORY MUSCLES
page 56

3-SECOND BIOGRAPHY
SIR RICHARD OWEN
1804–1892
An English naturalist and anatomist who showed how vertebrae in the early hominids' spine allowed our early ancestors to walk on two legs

30-SECOND TEXT
Judith Barbaro-Brown

The rib cage, or thoracic cage, consists of the sternum, the costal cartilages that connect it to the ribs, the 12 paired ribs, and the 12 thoracic vertebrae.

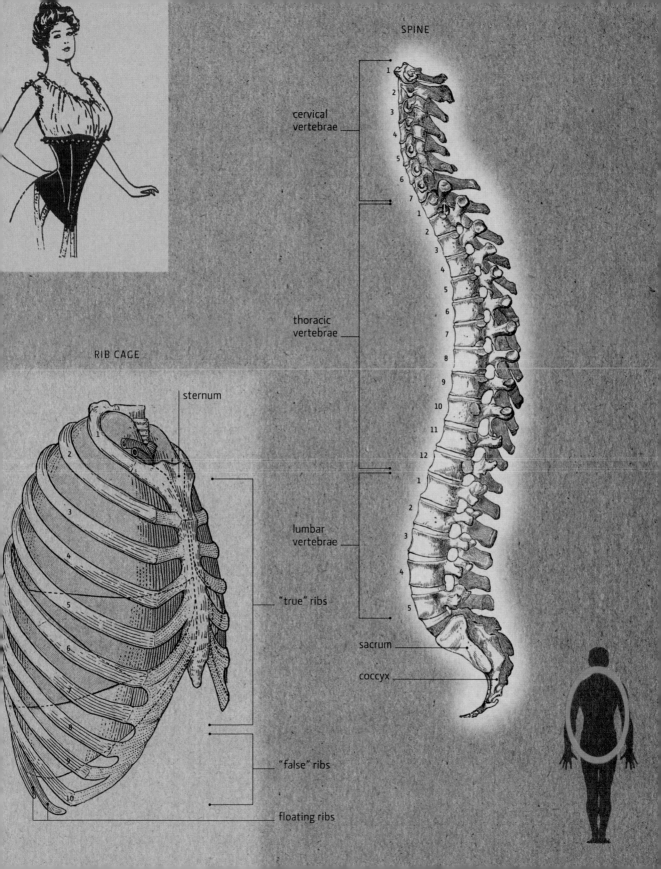

SPINE

cervical
vertebrae

1
2
3
4
5
6
7

thoracic
vertebrae

1
2
3
4
5
6
7
8
9
10
11
12

lumbar
vertebrae

1
2
3
4
5

sacrum

coccyx

RIB CAGE

sternum

1
2
3
4
5
6
7
8
9
10

"true" ribs

"false" ribs

floating ribs

1514
Born in Brussels (then part of the Habsburg Netherlands)

1528
Enrolled at University of Leuven/Louvain to study arts

1533
Moved to Paris to study medicine at the university, where his interest in anatomy was awakened

1536
Driven from Paris by political situation, and returned to Leuven, then moved on to Padua

1537
Graduated from Padua and became Professor of Surgery and Anatomy there

1538
Published anatomical drawings *Tabulae Anatomicae Sex*

1539
Began dissecting human corpses

1539
Updated Galen's handbook *Institutiones Anatomica*

1541
Published a corrected version of Galen's *Opera Omnia* and started writing his own anatomy book

1543
Public dissection of Jakob Carrer von Gebweiler, a notorious felon

1543
Publication of *De humani corporis fabrica* ("On the Structure of the Human Body")

1543
Appointed Imperial Physician at the court of Holy Roman Emperor Charles V

1544
Married Anne van Hamme

1555
Revised edition of *De humani* published

1564
Set off on a pilgrimage to the Holy Land, but was shipwrecked on the island of Zakynthos

1564
Died on Zakynthos

VESALIUS

The last and best known of a long line of distinguished physicians from what is now Belgium, Vesalius (Andries van Wesel, Andreas Vesal, or Andre Vesale) served at the court of the Holy Roman Emperor. After taking a medical degree at the University of Padua, Vesalius taught surgery and anatomy there and was later appointed Imperial Physician to Emperor Charles V, to whom he dedicated his magnum opus, the seven-volume *De Humani Corporis Fabrica* ("On the Structure of the Human Body"), 1543.

Vesalius is known as the father of modern human anatomy, but probably his greatest gift to medical science was his revolutionary—almost Holmesian—methods of teaching and research. He emphasized a hands-on approach, with rigorous observation, analysis, and checking, and the constant adjustment of theory to fit the facts, rather than the suppression of facts to fit established, impregnable theory. Before Vesalius, an anatomy lesson consisted of the students reading Galen's texts while a lackey under instructions from the teacher cut up the body. Students of Vesalius did their own cutting up (from 1539 on the bodies of hanged felons)

and took their own notes to compare with Galen. Using this method, Vesalius proved that Galen had based his entire theory on the dissection of barbary apes rather than human bodies, and that consequently some of his conclusions were wrong; then he outraged the establishment by not only publishing a corrected version of Galen's own works, but also producing a groundbreaking work of his own. What's more, Vesalius commissioned trained artists, probably from the studio of the Italian painter Titian, to record the dissection of bodies, bone, and muscle in immaculate detail and proportion. For this, he was roundly condemned by the establishment—including his own medical school. Rather than admit that Galen might have been wrong, Vesalius' one-time teacher Jacob Sylvius claimed that the human body must have changed since Galen's time. Attempts were even made to prove that Vesalius's methods were blasphemous, but such charges were dismissed in 1551 by an inquiry set up by Charles V (who had always been a supporter). Vesalius died 13 years later, in 1564, after being shipwrecked on the Greek island of Zakynthos while on pilgrimage to the Holy Land.

THE PELVIS

the 30-second anatomy

The pelvis is made up of the sacrum and coccyx at the back, with the paired coxal bones at the sides and front. Each coxal bone comprises the ilium, ischium, and pubis, joining together within the socket of the hip joint, and fusing at around 16 years of age. The pelvis contains trabecular and cortical bone, making it both strong and light. The pelvic inlet is a space within the pelvis. In men, it is heart-shaped, but in women, it is more oval and usually bigger, so there is more room for childbirth—the pelvis is also wider in women in order to accommodate a growing fetus. Pelvic ligaments are essential in providing stability, bridging the sacroiliac joints at the back, linking the lumbar vertebrae to the pelvis, and joining the pubic bones at the front. These ligaments are extremely strong because they must stabilize the weight of the upper body. The pelvic floor is made up of a number of intersupporting muscles that provide support for the pelvic and abdominal organs, and, by acting as a sphincter for the urethra, help maintain urinary continence.

3-SECOND INCISION
The pelvis is a ring of bone that contains and protects many organs, while supporting the upper body and connecting it to the lower limbs.

3-MINUTE DISSECTION
The normal aging process can have catastrophic effects on the pelvis. Significant wear and tear can occur at the sacroiliac joints; this lets pressure form on the spinal nerves running across this area, and leads to pain and discomfort across the buttocks and lower limbs, in a condition known as sciatica. Osteoporosis—thinning and loss of bone density—can also make it difficult for the pelvis to support the body weight, causing back pain.

RELATED TOPICS
See also
TYPES OF BONE TISSUE
page 16
THE BONE JOINTS
page 18
THE LOWER LIMBS
page 30
MOVEMENTS
page 42
THE PELVIC FLOOR MUSCLES
page 146

3-SECOND BIOGRAPHY
SIR JOHN CHARNLEY
1911–1982
An English orthopedic surgeon who pioneered joint replacement technology and designed an artificial hip joint in the mid-1960s

30-SECOND TEXT
Judith Barbaro-Brown

The left and right pubis bones meet in the front part of the pelvis and serve to protect the bladder.

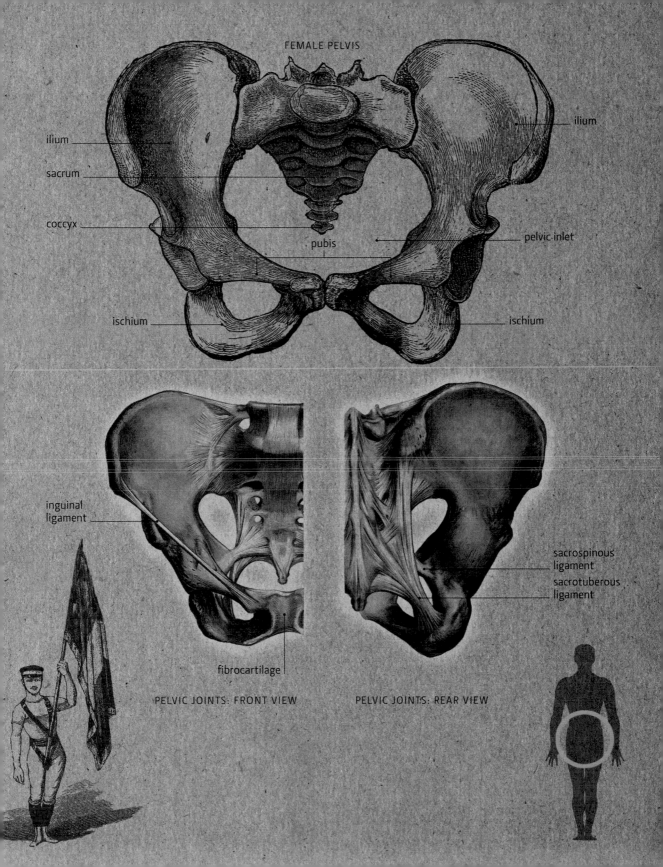

FEMALE PELVIS

ilium

ilium

sacrum

coccyx

pubis

pelvic inlet

ischium

ischium

inguinal
ligament

sacrospinous
ligament

sacrotuberous
ligament

fibrocartilage

PELVIC JOINTS: FRONT VIEW

PELVIC JOINTS: REAR VIEW

THE LOWER LIMBS

the 30-second anatomy

3-SECOND INCISION
The lower limb bones support the body and provide a strong and highly adaptive mechanism for walking.

3-MINUTE DISSECTION
The human lower limb is specially adapted to allow for walking on two feet. The bones are shaped, elongated, and very strong to support the heavy body while also acting as levers under the control of the powerful limb muscles. In addition, they have a mechanical function, transmitting kinetic motion from the hip joint to the ankle joint, letting walkers propel themselves forward and adapt to any terrain.

The lower limb begins at the inguinal ligament at the top of the pelvis, and ends at the talus in the ankle. The bones of the lower limb are the femur, patella, tibia, fibula, and talus. They are a mixture of long and sesamoid bones, containing both trabecular and cortical bone. The rounded, angled head of the femur makes up part of the hip joint; the base of the femur forms the knee joint with the tibia, fibula, and patella. The knee joint is the biggest synovial joint in the body, and—unlike the hip joint—is relatively unstable. To help improve this, ligaments hold together the femur, tibia, and fibula, with the patella (kneecap) held within the largest ligament on the front on the knee. The tibia is the main bone of the leg below the knee, with the much thinner fibula positioned laterally toward the outside of the leg. At their bases, the tibia and fibula form a bony arch, held together by ligaments, into which fits the rounded head of the talus to form the ankle joint. The talus acts to convert movement in the foot to movement in the hip, and vice versa.

RELATED TOPICS
See also
TYPES OF BONE TISSUE
page 16
THE BONE JOINTS
page 18
THE LIGAMENTS, CARTILAGE & TENDONS
page 20
MOVEMENTS
page 42
THE LOWER LIMB MUSCLE GROUPS
page 50

30-SECOND TEXT
Judith Barbaro-Brown

The upper section of the femur is at an angle of 120 degrees to the main part of the bone. The two femurs bear the whole weight of the upper body.

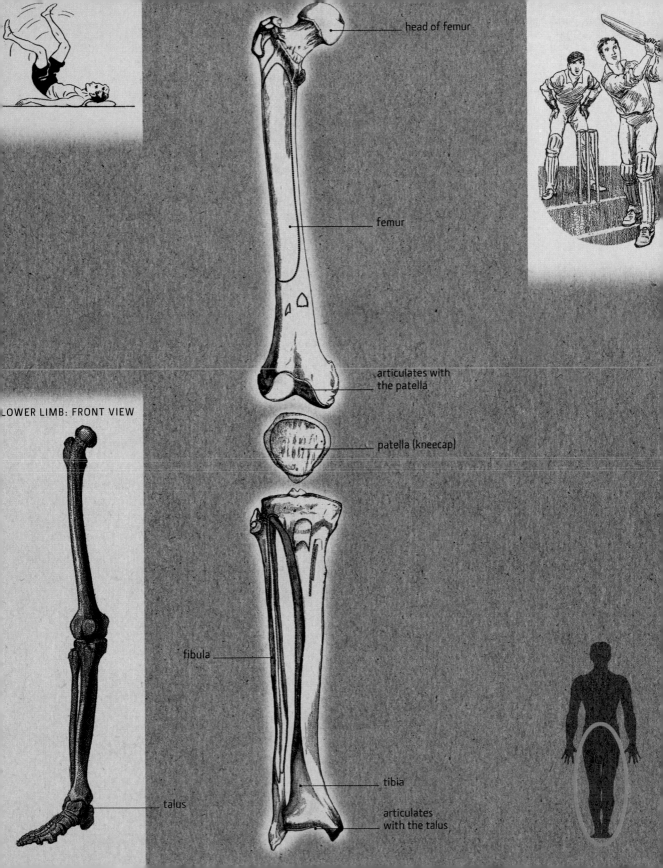

head of femur

femur

articulates with the patella

patella (kneecap)

LOWER LIMB: FRONT VIEW

fibula

tibia

talus

articulates with the talus

THE UPPER LIMBS

the 30-second anatomy

3-SECOND INCISION
The upper limb provides a flexible mechanism that lets humans carry large, heavy objects, but also facilitates intricate manipulation.

3-MINUTE DISSECTION
In a dislocated shoulder, the head of the humerus separates from the scapula; partial dislocation is called subluxation. Both dislocation and subluxation require realignment of the joint to prevent damage to the blood vessels and nerves in the area. The most common injuries at the elbow are caused by overuse, and are known as "tennis elbow" and "golfer's elbow," both painful conditions that are caused by excessive pulling on the humerus by the forearm tendons.

The upper limb begins at the shoulder, and includes the clavicle and scapula, forming the shoulder joint in combination with the humerus. This modified ball-and-socket joint is a little like the hip, but because the socket part of the shoulder is shallow, the head of the humerus is not held very tightly, making the shoulder joint highly mobile—but also unstable. At the elbow joint, the humerus articulates with the radius and ulna, which make up the forearm. The olecranon prominence at the tip of the elbow is formed by the base of the ulna. The soft area on the inner aspect of the elbow joint is called the antecubital fossa. It is here that a stethoscope is placed when measuring blood pressure, because the large brachial artery and vein can be found just below the skin; this also means that it is easy to access the vein to take a sample of blood. The arrangement of the radius and ulna, together with the three parts of the elbow joint, permits a large degree of rotation in the forearm. The forearm connects to the hand at the wrist, where the radius articulates with the scaphoid, lunate, and triquetral bones, but the ulna does not have a direct articulation with these bones.

RELATED TOPICS
See also
THE BONE JOINTS
page 18
THE LIGAMENTS, CARTILAGE & TENDONS
page 20
THE LOWER LIMBS
page 30
THE HANDS & FEET
page 34
MOVEMENTS
page 42

30-SECOND TEXT
Judith Barbaro-Brown

The humerus is typically around 14 inches (36 cm) long. At its base two epicondyles (knucklelike areas) can be felt, one on each side of the elbow.

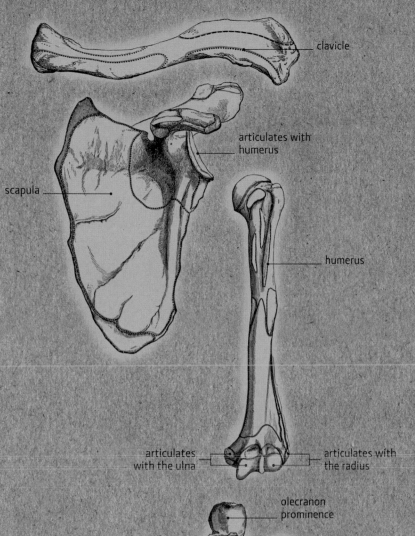

clavicle

articulates with
humerus

scapula

humerus

articulates
with the ulna

articulates with
the radius

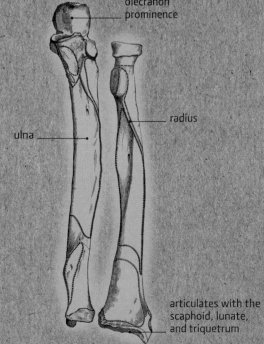

olecranon
prominence

ulna

radius

articulates with the
scaphoid, lunate,
and triquetrum

UPPER LIMB:
FRONT VIEW

THE HANDS & FEET

the 30-second anatomy

3-SECOND INCISION
The bones of the hand are adapted to provide manual dexterity and those in the feet to provide a stable platform for supporting the body and walking.

3-MINUTE DISSECTION
A man's average hand length is 7⅖ inches (18.9 cm), while the average for a woman is 6⅘ inches (17.2 cm). The hand begins to form in the seventh week of pregnancy, the foot in the eighth week. By week 12, hands and feet are formed, with fingers and toes. Footprints and fingerprints have formed by week 24, and nails have reached the ends of the digits by week 35.

Each hand contains 27 bones— 8 in the wrist, 5 in the palm, and 14 in the fingers and thumb. Human hands have an opposable thumb, allowing for manipulation of small, delicate objects as well as large, heavy ones. The hand has an excellent nerve supply, with the highest density of nerve endings occurring in the fingertips. The left hand is controlled by the right side of the brain, and vice versa—the preference of hand for writing indicates the dominant-functioning side of the brain. Each foot contains 28 bones—2 in the rearfoot, 5 in the midfoot, 19 in the forefoot, and 2 sesamoid bones sitting in a tendon beneath the big toe joint. The toes and midfoot bones are aligned not for grasping objects but to provide a stable platform to support the body. Articulations in the foot let it adapt to uneven ground when walking, and provide a rigid lever to propel the body forward during walking; the hand, in contrast, does not have to support body weight and so is directly adapted to give maximum flexibility. Hands and feet are controlled by groups of muscles originating in the arms and legs respectively, together with smaller muscles within each hand and foot.

RELATED TOPICS
See also
TYPES OF BONE TISSUE
page 16
THE LOWER LIMBS
page 30
THE UPPER LIMBS
page 32
TYPES OF MUSCLE TISSUE
page 40

3-SECOND BIOGRAPHY
WILHELM RÖNTGEN
1845–1923
A German physicist and the first person to make what we now call an X-ray (of his wife's hand) using a cathode ray tube

30-SECOND TEXT
Judith Barbaro-Brown

Each thumb and each big toe contains two phalanges, while each finger and each of the other toes has three—making a total of 56 phalanges.

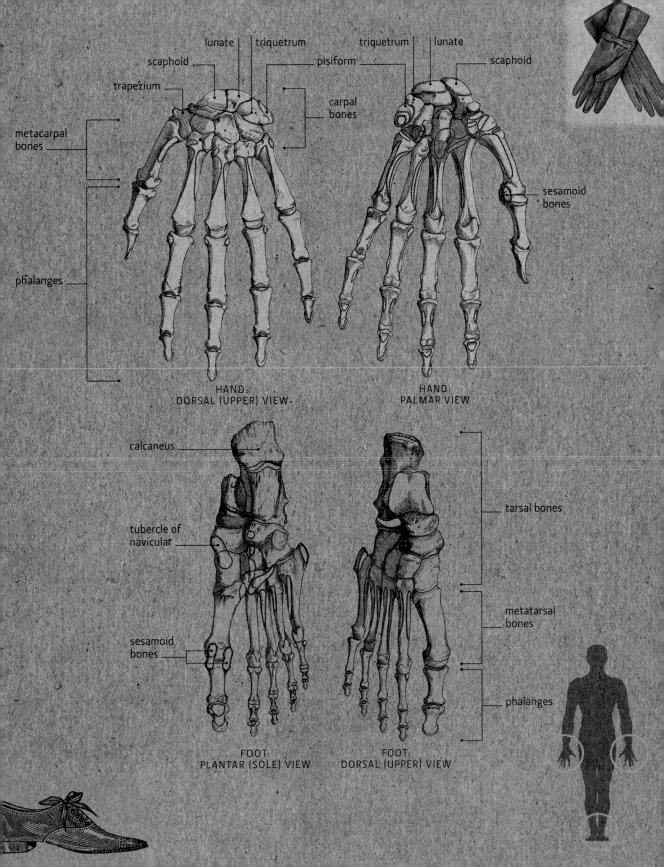

metacarpal
bones

phalanges

trapezium

scaphoid

lunate

triquetrum

pisiform

carpal
bones

triquetrum

lunate

scaphoid

sesamoid
bones

HAND:
DORSAL (UPPER) VIEW.

HAND:
PALMAR VIEW

calcaneus

tubercle of
navicular

sesamoid
bones

tarsal bones

metatarsal
bones

phalanges

FOOT:
PLANTAR (SOLE) VIEW

FOOT:
DORSAL (UPPER) VIEW

THE MUSCULAR SYSTEM

abductor/adductor muscles that work in pairs to move a body part; the abductor moves a body part away from the body and its central axis. The opposite movements, toward the body or across its central axis, are brought about by adductor muscles.

anterior/posterior Front/back of the body. An anterior muscle is on the front section of a limb, while its posterior counterpart is on the back section.

biceps brachii Muscle on the front (anterior) of the upper arm whose main function is as a flexor—to bend the arm at the elbow joint. The biceps brachii is balanced by the triceps brachii on the rear (posterior) of the upper arm; its main function is as an extensor—to extend or straighten the arm. A biceps muscle has two heads (is attached at two points at its upper part), while the triceps has its origin in three heads (or attachments in its upper part). Another notable biceps muscle is the biceps femoris in the rear of the thigh; this serves to bend the knee joint and to straighten the hip joint.

cardiac muscle One of three types of muscles in the body, with skeletal muscle and smooth muscle. Cardiac muscle is in the myocardium (the heart's muscular layer) and contracts rhythmically throughout a person's life.

constrictor Muscle that contracts or compresses a body part—for example, one that closes a bodily orifice. Muscles that open and close orifices are also called sphincters.

depressor/levator Muscles of lowering and raising. A depressor muscle lowers a part of the body, while a levator muscle raises it.

diaphragm Very thin, dome-shaped structure containing tendon tissue as well as muscle that separates the body's thoracic cavity (chest) from the abdominal cavity. The diaphragm plays a key role in breathing. When it contracts and flattens, the volume of the thoracic cavity increases significantly, then when it relaxes and rises the volume is reduced; at the same time intercostal muscles lift and lower the rib cage.

extensor/flexor Muscles of straightening and bending. Extensors increase the angle between bones at a joint; flexors decrease this angle. For example, extensor muscles straighten the leg by increasing the angle between bones of the lower and upper leg at the knee joint; flexor muscles bend the arm by decreasing the angle between the bones of the forearm and the upper arm at the elbow.

gluteal muscles The muscles in the buttocks; there are nine gluteal muscles in each buttock. The gluteus maximus is the largest and gives the buttock its rounded form, and, together with the gluteus medius and gluteus minimus, it is among the three nearest the body's surface. They serve to straighten the thigh, also moving the femur outward and rotating it at the hip joint.

pronator/supinator Muscles of rotation. The pronator rotates the forearm so that the palm faces backward or down, while the supinator rotates it so the palm faces forward or up.

skeletal muscle One of three types of muscles in the body, with cardiac muscle and smooth muscle. The skeletal muscles are controlled by the nervous system—each skeletal muscle is connected by nerves to the spinal cord and brain. The body uses skeletal muscles to make conscious movements, such as lifting the leg or bending the arm; skeletal muscle also works under subconscious control when, for example, it supports the head and limbs of a person who is standing upright.

smooth muscle One of three types of muscles in the body, with cardiac muscle and skeletal muscle. Smooth muscles carry out involuntary bodily actions, such as in the stomach and intestines, or the urinary bladder. Peristalsis—the series of rhythmic contractions that move food through the gastrointestinal tract—is performed by smooth muscles.

TYPES OF
MUSCLE TISSUE

the 30-second anatomy

3-SECOND INCISION
There are three muscle types: skeletal (in muscles that move the skeleton), cardiac (in the heart), and smooth (in lungs, blood vessels, organs, and gut).

3-MINUTE DISSECTION
Skeletal muscles are usually under voluntary control—they are controlled consciously. Cardiac and smooth muscle are controlled involuntarily, and their actions are brought about by the autonomic nervous system, outside of our conscious control. As a result, heart rate, blood pressure, blood flow, and breathing can change very quickly to respond to the environment, helping the body to function well in stressful situations.

Skeletal muscle is also called striated muscle because alternating dark and light bands (striations) are visible along the length of its cell fibers when these are viewed under the microscope. These bands occur because the proteins making up the muscle are in the form of thick and thin protein filaments that overlap one other. The thick filaments are myosin; the thin filaments consist of three proteins—actin, tropomyosin, and troponin. The dark band corresponds to the thick filaments overlapping at each end with thin filaments; the light band is where there are only thin filaments. The muscle contracts when the thick and thin filaments slide over each other, making the muscle shorter. Cardiac muscle cells work in the same way, but are smaller than skeletal muscle cells, and they are joined together by specialized plates called intercalcated disks, which help the cardiac muscle cells contract together at the same time. Actin and myosin are present in all three muscle types. In skeletal and cardiac muscle cells, these proteins are organized in units called sarcomeres, with thin and thick filaments. Smooth muscle cells are not arranged in sarcomeres and have no striations. They still contain actin and myosin, but the proteins are arranged multidirectionally, so contraction occurs in all directions.

RELATED TOPICS
See also
MOVEMENTS
page 42
THE CIRCULATORY SYSTEM
page 62
THE HEART
page 64
THE LUNGS
page 76

3-SECOND BIOGRAPHY
GUILLAUME DUCHENNE
1806–1875
A French neurologist and the first clinician to practice muscle biopsy. His name is linked to several diseases including Duchenne muscular dystrophy

30-SECOND TEXT
Judith Barbaro-Brown

Cardiac muscle works without ceasing—it never grows tired. By the time a person is 70, it will have contracted 2,500 million times.

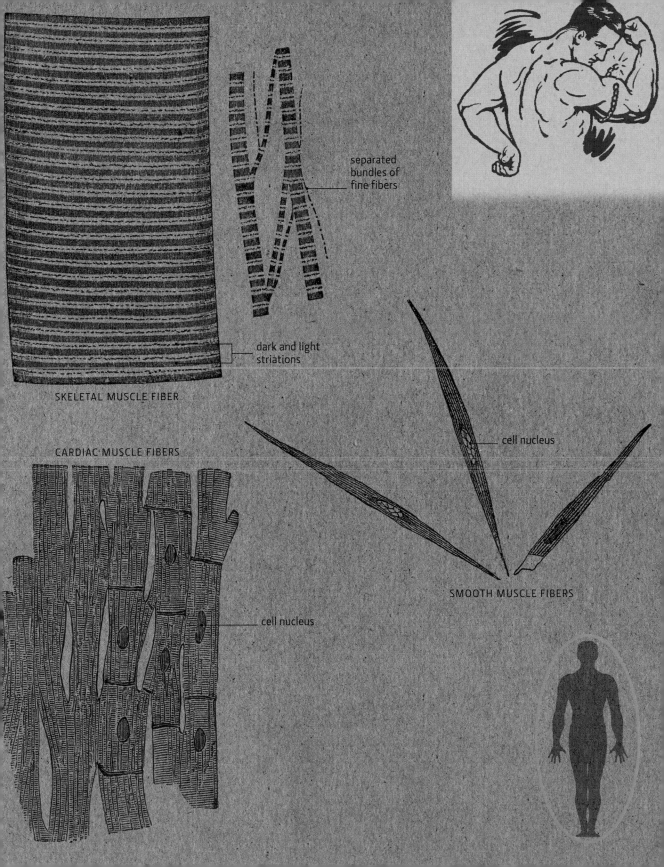

SKELETAL MUSCLE FIBER

separated bundles of fine fibers

dark and light striations

CARDIAC MUSCLE FIBERS

cell nucleus

SMOOTH MUSCLE FIBERS

cell nucleus

MOVEMENTS

the 30-second anatomy

3-SECOND INCISION
People are able to move in several dimensional planes, and combinations of these movements provide a high degree of flexibility.

3-MINUTE DISSECTION
Some individuals have increased movement and mobility in the musculoskeletal system; this is known as hypermobility. For example, some people can put one leg behind the head, while others can bend their knee joints forward as well as back.

The musculoskeletal system's

three planes of movement are sagittal, frontal (coronal), and transverse, with each plane a division through the body. The sagittal plane passes from front to back, forming left and right sections; it allows for forward and backward movement, known as flexion and extension. The frontal plane passes from one side to the other, splitting the body into front and back; it allows for side-to-side movements called abduction and adduction. The transverse plane splits the body into top and bottom, and movements in this plane take the form of rotation. Movements at joints are categorized in relation to these planes: uniplanar joints (such as interphalangeal joints in the fingers) allow for movement in only one plane; biplanar joints (such as the atlantoaxial joint in the neck) have movement in any two planes; and multiplanar joints (such as the shoulder joint) allow for movement in all three planes simultaneously. Muscles and tendons facilitate movement, but to prevent damage to the skeletal system, ligaments limit movement.

RELATED TOPICS
See also
THE BONE JOINTS
page 18
THE LIGAMENTS, CARTILAGE & TENDONS
page 20

30-SECOND TEXT
Judith Barbaro-Brown

Muscles can work in a single movement plane, such as when nodding the head. They can also work in all three planes of movement, such as when rotating the head in a circular motion.

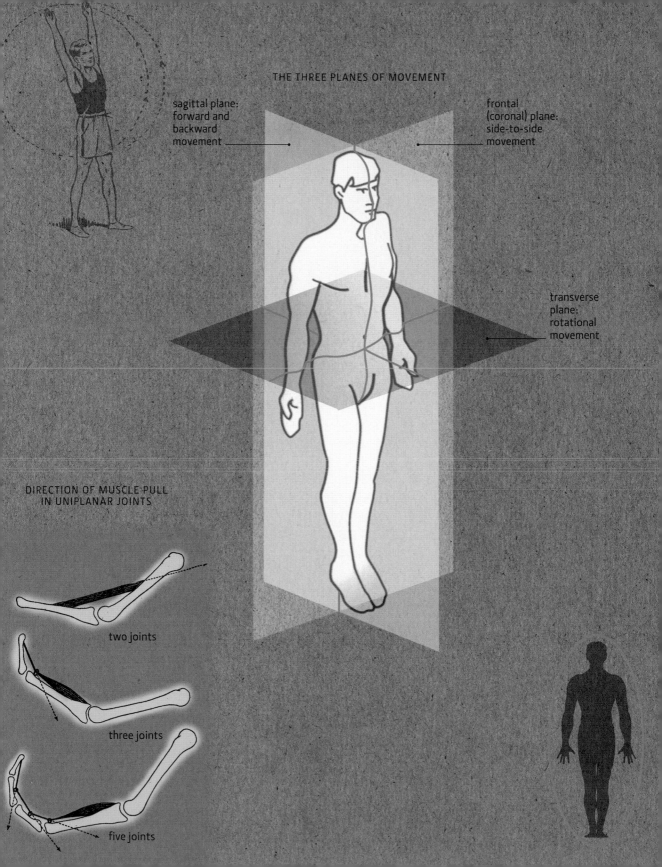

THE THREE PLANES OF MOVEMENT

sagittal plane:
forward and
backward
movement

frontal
(coronal) plane:
side-to-side
movement

transverse
plane:
rotational
movement

DIRECTION OF MUSCLE PULL
IN UNIPLANAR JOINTS

two joints

three joints

five joints

THE FACIAL MUSCLES

the 30-second anatomy

3-SECOND INCISION
There are more than
20 facial muscles—they
let people convey their
emotions and control the
opening or closing of
the eyes and mouth.

3-MINUTE DISSECTION
Illness can cause damage
to the facial nerve and
paralysis of the facial
muscles, which droop; the
damage can be permanent
or temporary. When the
facial nerve damage has no
known cause, the paralysis
is known as Bell's palsy.
Usually it affects one side
of the face. The paralysis
can last from a few days to
a couple of years.

Winking, puffing the cheeks out, or pouting with the lips are all actions that use the facial muscles. Muscles of facial expression arise from the facial bones of the skull. These muscles move when stimulated by the facial nerve; impulses from the facial nerve cause the muscle fibers to contract and this brings about their movement. Facial muscles are classified by their primary action: muscles that dilate a structure, such as those that flare out the nostrils, are known as dilators; muscles that enable facial expressions, such as those that raise the eyebrows, are known as expressors; and muscles that open or close structures, such as those that close the eyes or mouth, are known as sphincteric muscles. The names of the facial muscles indicate their actions—levators lift their associated structures, depressors pull down their associated structures, and corrugators wrinkle the overlying skin as well as moving the associated muscles. The fibers that make up the muscles insert into a layer of the skin (fascia). When the muscle fibers contract, the overlying skin moves. For example, when the muscle around the eyes (orbicularis oculi) contracts, the eyelids close, acting like a sphincter around the eyeball. The ability to close the eyelids protects the eyeballs from bright light and injury. Facial muscles control the degree of opening and shape of the mouth, which is vital for speech.

RELATED TOPICS
See also
THE SKULL
page 22
TYPES OF MUSCLE TISSUE
page 40

3-SECOND BIOGRAPHY
CHARLES BELL
1774–1842
A Scottish anatomist and
surgeon, the discoverer of
Bell's palsy

30-SECOND TEXT
Gabrielle M. Finn

The mentalis muscle in the chin lifts the lower lip and as a result is sometimes called "the pouting muscle."

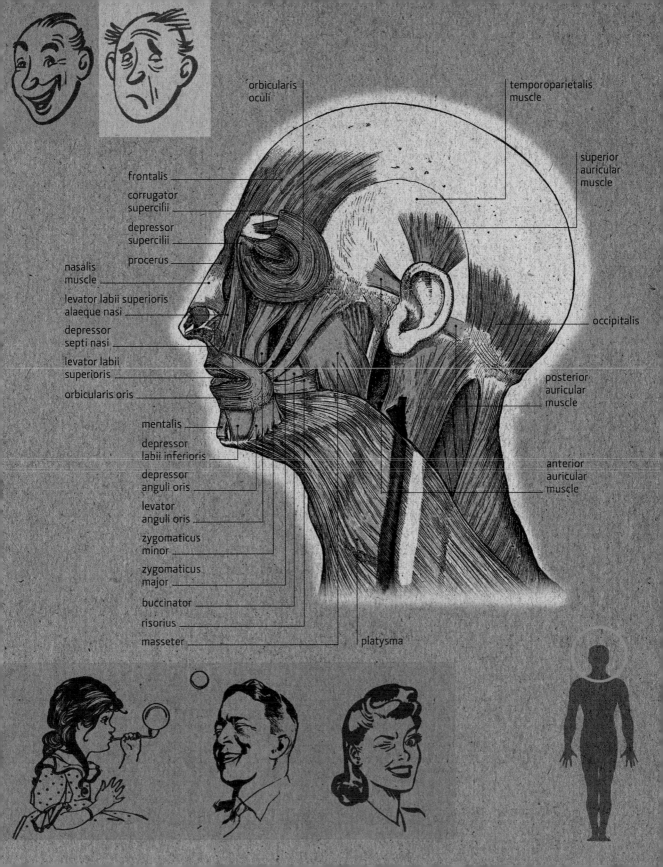

orbicularis
oculi

temporoparietalis
muscle

frontalis

superior
auricular
muscle

corrugator
supercilii

depressor
supercilii

procerus

nasalis
muscle

levator labii superioris
alaeque nasi

occipitalis

depressor
septi nasi

levator labii
superioris

orbicularis oris

posterior
auricular
muscle

mentalis

depressor
labii inferioris

depressor
anguli oris

anterior
auricular
muscle

levator
anguli oris

zygomaticus
minor

zygomaticus
major

buccinator

risorius

masseter

platysma

THE NECK MUSCLES

the 30-second anatomy

The neck is a fibrous tube that runs between the head and thorax. It contains four other compartments: the vertebral, visceral, and two carotid compartments, one on each side. Other structures that traverse the neck are the esophagus, trachea, internal carotid artery, and internal jugular vein. On each side, the sternocleidomastoid muscle divides the neck into a rear (posterior) and front (anterior) triangle, with eight muscles in each. The rear muscles are used to nod the head. Four (rectus capitis anterior, rectus capitis lateralis, longus colli, and longus capitis) are attached to bones on the vertebral column, while the other four (scalenus anterior, scalenus medius, scalenus posterior, and levator scapula) attach to limb bones. Of the eight front muscles, four "suprahyoid muscles" lie above and help to raise the hyoid bone (between the chin and thyroid cartilage). These muscles—geniohyoid, stylohyoid, digastric, and mylohyoid—also pull the hyoid bone forward and backward, raise the floor of the mouth, or lower the mandible when chewing and swallowing. "Infrahyoid muscles" beneath the hyoid are the sternohyoid, thyrohyoid, omohyoid, and sternothyroid; they depress the hyoid bone or the thyroid cartilage and help in breathing and speech.

3-SECOND INCISION
Neck muscles enable a person to nod the head, swallow, and breathe.

3-MINUTE DISSECTION
When awake, a person is consciously aware of the action of swallowing because of its complexity, yet in sleep a normal adult swallows on average 3–7 times per hour without knowing it. Overall, we swallow 25 or more times in a night—and still wake up hungry!

RELATED TOPICS
See also
THE FACIAL MUSCLES
page 44

3-SECOND BIOGRAPHY
PIERRE AUGUSTIN BÉCLARD
1785–1825
A French anatomist who described one of the minor triangles of the neck that contains the lingual artery and the hypoglossal nerve

30-SECOND TEXT
December S. K. Ikah

The main muscles used for flexing (bending) the neck are the sternocleidomastoid and the trapezius.

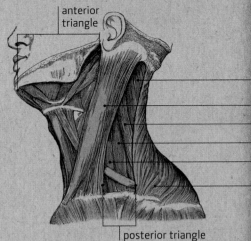

anterior triangle

splenius capitis

sternocleidomastoid

levator scapulae

scalenus medius

scalenus anterior

trapezius

posterior triangle

SUPERFICIAL MUSCLES: SIDE VIEW

geniohyoid

digastric

mylohyoid

stylohyoid

hyoid bone

thyrohyoid

omohyoid

sternohyoid

sternothyroid

SUPERFICIAL MUSCLES: FRONT VIEW

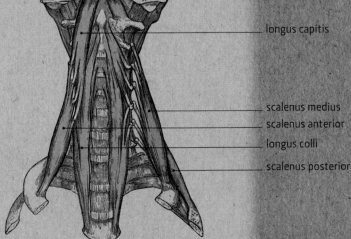

longus capitis

scalenus medius

scalenus anterior

longus colli

scalenus posterior

DEEP MUSCLES: FRONT VIEW

THE UPPER LIMB MUSCLE GROUPS

the 30-second anatomy

3-SECOND INCISION
The upper limb muscles are found between the shoulder and the tips of the fingers—they act to bring about a wide range of movements.

3-MINUTE DISSECTION
Have you ever wondered what the carpal tunnel is? Imagine an electrical cable—it contains several wires and is encased within a plastic coat. The carpal tunnel is similar in construction and is the point at which blood vessels, nerves, and muscle tendons cross the wrist. These structures can become compressed between the tunnel roof (a sheath of tough tissue) and floor (bone) or else the structures themselves can become inflamed within the tunnel.

The muscles of the upper limb

work like a lever system: Muscle movement brings about an action at the neighboring joint. The muscles that run from the shoulder to the elbow (biceps brachii) on the front (anterior) surface of the upper arm work primarily to flex (bend) the arm at the elbow joint. The muscles on the rear (posterior) surface of the upper arm (triceps brachii) straighten (extend) the arm at the elbow joint. The biceps and triceps are powerful muscles, separated by the humerus (the long bone of the upper arm). Muscles that cross the elbow joint act on the joint, forearm, wrist, or fingers. The general rule for the muscles of the forearm is that flexors are on the front and extensors on the rear; those muscles on the front are typically larger and stronger than those on the rear. Muscles of the forearm bring about movements, including bending (flexion), straightening (extension), moving away from or across the body's central axis (abduction or adduction), rotating so the palm faces backward or down (pronation), and rotating so the palm faces forward or up (supination). All muscles of the upper limb are supplied by nerves arising from the brachial plexus.

RELATED TOPICS
See also
THE UPPER LIMBS
page 32
THE HANDS & FEET
page 34
THE NERVE PLEXUSES
page 138

30-SECOND TEXT
Gabrielle M. Finn

The biceps brachii in the front of the upper arm is the muscle flexed to an impressive bulge by bodybuilders.

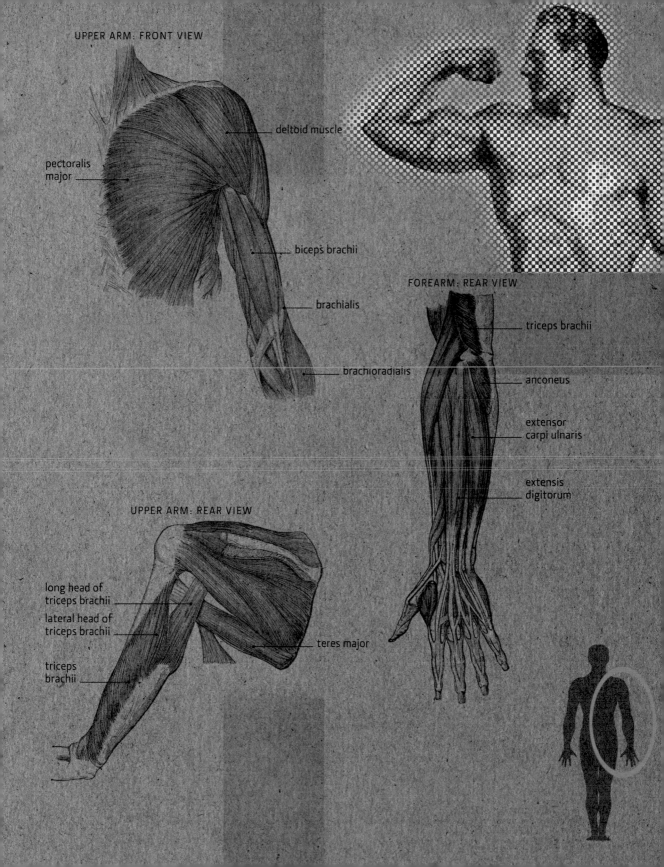

UPPER ARM: FRONT VIEW

deltoid muscle

pectoralis major

biceps brachii

brachialis

brachioradialis

FOREARM: REAR VIEW

triceps brachii

anconeus

extensor carpi ulnaris

extensis digitorum

UPPER ARM: REAR VIEW

long head of triceps brachii

lateral head of triceps brachii

triceps brachii

teres major

THE LOWER LIMB MUSCLE GROUPS
the 30-second anatomy

3-SECOND INCISION
Lower limb muscles are
organized into muscles of
the thigh, gluteal region
(buttocks area), and leg
(anatomically, leg means
from knee to ankle).

3-MINUTE DISSECTION
Muscular injuries within
the lower limb are a
common side effect of
playing sports. The thigh
muscles are frequently
injured during activities
that involve running, such
as soccer or basketball.
The force of muscle
exertion can tear the
muscle fibers, or even in
extreme cases tear the
muscles from their bony
origins and insertions.
Such injuries can result
from inadequate warm-up
or from an external force,
such as being kicked by
another player.

In the front (anterior) thigh are
two major muscle groups—the hip flexors and
the extensors of the knee. The hip flexors—
pectineus, sartorius, and iliopsoas—bring about
bending (flexion) of the femur at the hip joint;
these muscles also assist in rotating the thigh.
The extensors in the front (anterior) thigh are
called the quadriceps femoris and comprise
rectus femoris, vastus lateralis, vastus medialis,
and vastus intermedius; these extend the leg
at the knee joint and help with flexion of the
thigh at the hip joint. The antagonists (opposite
muscles) to the quadriceps are the hamstrings in
the rear (posterior) of the thigh: biceps femoris,
semitendinosus, and semimembranosus. These
muscles primarily bend and rotate the leg. The
gluteal muscles within the buttocks also extend
the thigh, move the femur outward (abduction),
and rotate the femur at the hip joint. Below the
knee joint are several leg muscles. Those in the
front region (anterior compartment) dorsiflex
the foot, meaning that the heel is lowered to
the floor and the toes point upward, as seen
during walking; those in the leg's rear section
(posterior compartment) plantarflex the foot—
this is when the toes are depressed and the
heel rises from the floor.

RELATED TOPICS
See also
TYPES OF BONE TISSUE
page 16
THE BONE JOINTS
page 18
THE LOWER LIMBS
page 30
THE HANDS & FEET
page 34
MOVEMENTS
page 42

30-SECOND TEXT
Gabrielle M. Finn

*The tibialis anterior
and the extensor
hallucis longor
both play a role in
dorsiflexing the foot
(lifting the toes up).*

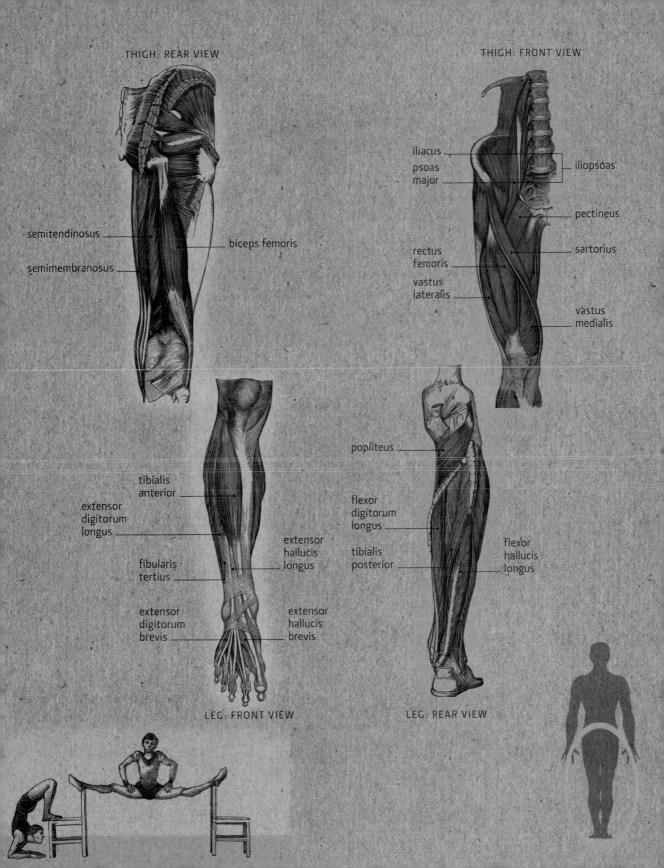

THIG: REAR VIEW

THIGH: FRONT VIEW

semitendinosus

semimembranosus

biceps femoris

iliacus

psoas major

iliopsoas

pectineus

rectus femoris

sartorius

vastus lateralis

vastus medialis

tibialis anterior

extensor digitorum longus

fibularis tertius

extensor digitorum brevis

extensor hallucis longus

extensor hallucis brevis

LEG: FRONT VIEW

popliteus

flexor digitorum longus

tibialis posterior

flexor hallucis longus

LEG: REAR VIEW

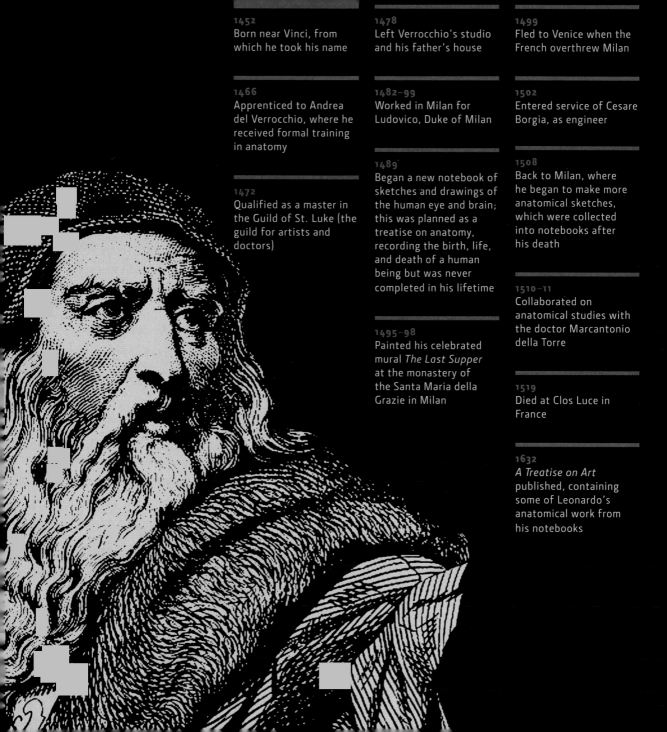

1452
Born near Vinci, from which he took his name

1466
Apprenticed to Andrea del Verrocchio, where he received formal training in anatomy

1472
Qualified as a master in the Guild of St. Luke (the guild for artists and doctors)

1478
Left Verrocchio's studio and his father's house

1482–99
Worked in Milan for Ludovico, Duke of Milan

1489
Began a new notebook of sketches and drawings of the human eye and brain; this was planned as a treatise on anatomy, recording the birth, life, and death of a human being but was never completed in his lifetime

1495–98
Painted his celebrated mural *The Last Supper* at the monastery of the Santa Maria della Grazie in Milan

1499
Fled to Venice when the French overthrew Milan

1502
Entered service of Cesare Borgia, as engineer

1508
Back to Milan, where he began to make more anatomical sketches, which were collected into notebooks after his death

1510–11
Collaborated on anatomical studies with the doctor Marcantonio della Torre

1519
Died at Clos Luce in France

1632
A Treatise on Art published, containing some of Leonardo's anatomical work from his notebooks

LEONARDO DA VINCI

It is a cliché that Leonardo da Vinci was the epitome of the Renaissance man, a polymath who ignored boundaries and combined the skills and talents of artist, engineer, musician, scientist, architect, inventor, cartographer, geologist, anatomist, and more. But clichés generally become clichés because they are true—Leonardo (or Leonardo di ser Piero da Vinci) was an extraordinary man, born in obscurity and out of wedlock in Vinci, a small town outside Florence, and dying, legend has it, in the arms of the king of France, his last patron. His talents were sought by kings and princes and he survived apparently unscathed the patronage of popes and the notorious Borgias. Nonetheless, after a lifetime of creativity and invention, he died regretting that he had never completed a single project.

Anatomical drawing was just one aspect of his prodigious output, and arose from his training in art. Leonardo was apprenticed to the Florentine artist Andrea del Verrocchio, who insisted that all his apprentices study anatomy. Leonardo learned that observation and maintaining a meticulous visual record were far more useful than long-winded verbal description. He became a master of topographical anatomy (recording visible muscles and sinews), eventually producing his iconic drawing *Vitruvian Man*, in 1490. As a qualified artist he was allowed to attend dissections of human bodies and worked at hospitals in Florence and Milan, collaborating with Doctor Marcantonio della Torre to produce finely detailed anatomical drawings of the inside of the human body, including the skeleton, heart, blood vessels, sexual organs, and a fetus in utero. Everything was recorded in meticulous detail, down to the tiniest capillary and from several angles, producing a priceless record for future anatomists. Leonardo assembled more than 200 pages of drawings and notes in preparation for a planned treatise on anatomy, but true to form did not finish the project. The papers were left to his pupil, Francesco Melzi, who was also defeated by them; finally, in 1632, 50 years after Melzi's own death and 113 after that of Leonardo, a few of the drawings were published in *A Treatise on Art*. The rest were preserved; their full importance was realized in the 18th century by William Hunter (see pages 148–149), who used them as a model for his own anatomical illustrations.

THE ABDOMINAL & BACK MUSCLES

the 30-second anatomy

3-SECOND INCISION
The abdominal and back muscles stabilize the vertebral column and support the spine and thoracic, abdominal, and pelvic organs.

3-MINUTE DISSECTION
Contraction of the muscles of the abdominal wall brings about an increase in pressure within the abdomen, which produces the force required for defecation, urination, vomiting, heavy lifting, and childbirth. Because the abdominal wall is weak, especially around the navel, increased pressure can sometimes cause a hernia, where abdominal content protrudes toward the skin. This often requires surgical intervention to reposition it back within the abdominal cavity.

The muscles of the abdomen and back maintain posture, move the trunk, or torso, and compress the contents of the abdominal cavity. The back muscles are responsible for supporting the spine and moving the spine, shoulders, arm, and neck. The muscles nearest the surface of the body (superficial muscles) include trapezius, latissimus dorsi, and the rhomboids; these are responsible for moving the appendicular skeleton (the pectoral girdle, pelvis, and limbs) and, in particular, the scapula (shoulder blade). The medium-depth (intermediate) back muscles—serratus anterior and posterior—assist in moving the ribs during breathing. The deep muscles of the back, including spinotransversales and erector spinae, move the head and extend the vertebral column. The abdominal wall has four flat muscles— external oblique, internal oblique, transverses abdominus, and rectus abdominis. These flex the trunk and compress the abdominal contents. Of these, rectus abdominis is nearest the surface. It is separated into right and left muscles by the linea alba (fibrous tissue) at the midline, and is divided also by horizontal fibrous bands called tendinous intersections.

RELATED TOPICS
See also
MOVEMENTS
page 42
THE RESPIRATORY MUSCLES
page 56

30-SECOND TEXT
Gabrielle M. Finn

On people with an athletic physique, tendinous intersections are visible; the resultant division in the rectus abdominis is what we call a six-pack.

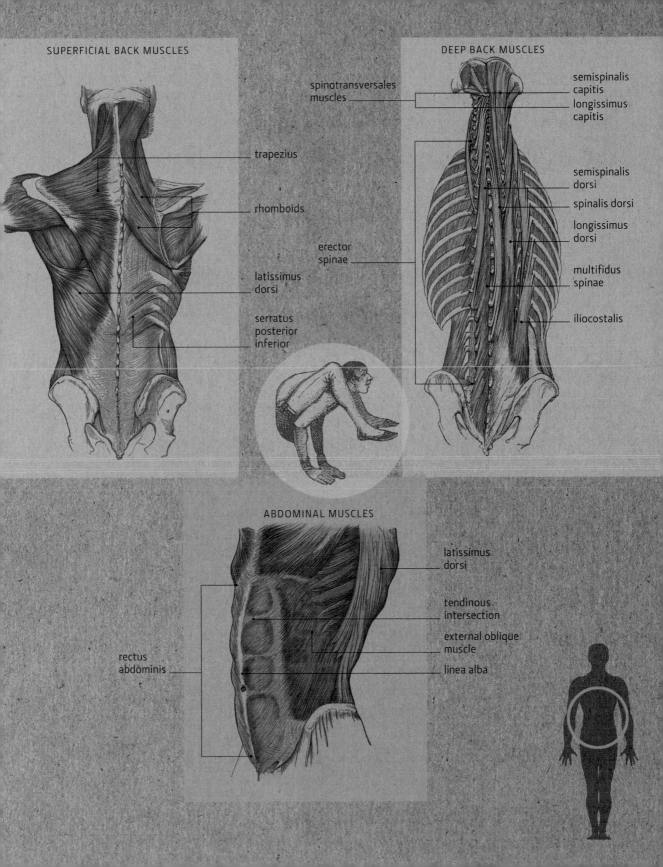

SUPERFICIAL BACK MUSCLES

trapezius

rhomboids

latissimus
dorsi

serratus
posterior
inferior

DEEP BACK MUSCLES

spinotransversales
muscles

semispinalis
capitis

longissimus
capitis

semispinalis
dorsi

spinalis dorsi

longissimus
dorsi

multifidus
spinae

iliocostalis

erector
spinae

ABDOMINAL MUSCLES

latissimus
dorsi

tendinous
intersection

external oblique
muscle

linea alba

rectus
abdominis

THE RESPIRATORY MUSCLES

the 30-second anatomy

The most important respiratory muscles are the diaphragm, and the external and internal intercostal muscles. The diaphragm is a thin, dome-shaped structure containing both muscle and tendon tissue, attached to the body wall and rib cage. It partitions the thoracic and abdominal cavities, and important structures pass through it, such as the esophagus, thoracic aorta, and inferior vena cava. During quiet breathing, the diaphragm contracts and flattens, its central portion is pulled down, and this increases the vertical volume of the thoracic cavity; when it relaxes and elevates, the vertical volume reduces. Eleven pairs of intercostal muscles run at different angles in two layers between the ribs. When a person is breathing in, external intercostals widen the rib cage by lifting the ribs upward and outward; when he or she is breathing out, the internal intercostals lower the ribs, narrowing the rib cage. The front end of a rib connects with the sternum or a neighboring rib and lies beneath its rear end, which connects with the vertebral column. When a rib is lifted, the sternum moves upward and forward, increasing the anteroposterior (front/back) dimension. Air is sucked into the lungs as the pressure in the thoracic cavity drops and the thorax expands.

3-SECOND INCISION
Respiratory muscles are essential for breathing — when they contract, they enable movement of the rib cage, so shifting air into and out of the lungs.

3-MINUTE DISSECTION
Thoracic movements can be felt by placing the palms on each side of the rib cage while breathing in and out. The increase in lateral dimension is like the movement of raising a bucket handle—the handle of the bucket (the ribs) moving upward and outward. Increase in anteroposterior dimension is analogous to lifting a pump handle—a forward and upward movement. The meat that diners enjoy on a sparerib is an animal's intercostal muscle!

RELATED TOPICS
See also
THE LUNGS
page 76

3-SECOND BIOGRAPHY
VESALIUS (ANDREAS VESAL)
1514–1564
A Flemish anatomist who noted the number of ribs in 1543

30-SECOND TEXT
Jo Bishop

The muscle fibers of the diaphragm are attached all around to the lower part of the chest and enclose an area of fibrous tissue shaped like a trefoil.

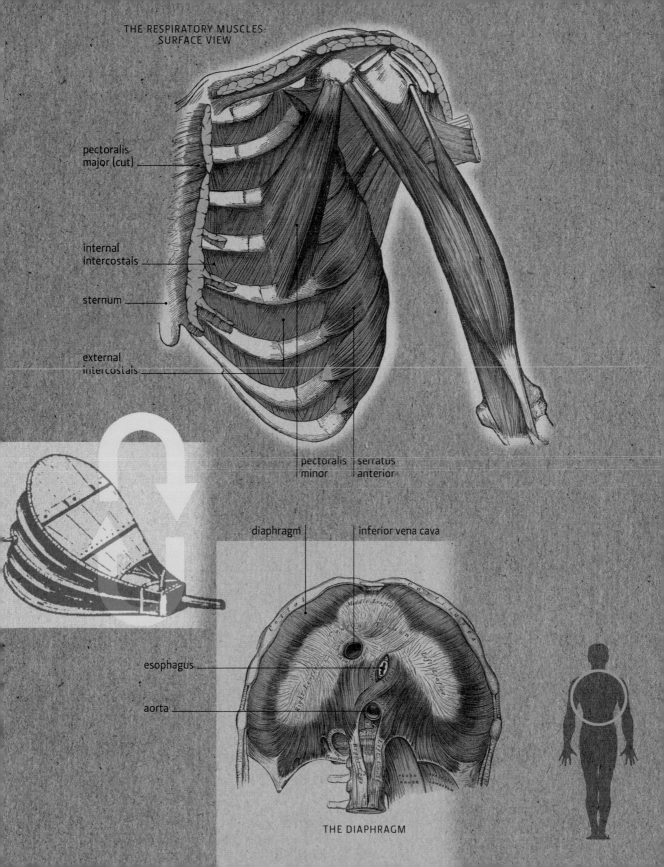

THE RESPIRATORY MUSCLES:
SURFACE VIEW

pectoralis
major (cut)

internal
intercostals

sternum

external
intercostals

pectoralis
minor

serratus
anterior

diaphragm

inferior vena cava

esophagus

aorta

THE DIAPHRAGM

THE CARDIOVASCULAR
& RESPIRATORY SYSTEMS

THE CARDIOVASCULAR
& RESPIRATORY SYSTEMS
GLOSSARY

atrium One of two upper chambers of the heart; the atria were previously known as the auricles.

capillaries Tiny blood vessels, part of the body's microcirculation. Through the walls of the capillaries oxygen and nutrients pass from the blood supply into body tissue, while carbon dioxide and waste products pass from the tissue into the blood.

cardiac cycle Set of heart muscle contractions in which blood is received from the veins, passes through the heart, and is pumped out into the arteries. Its two principal phases are the diastole (blood flows in) and the systole (blood is pushed first from the atria into the ventricles and then out into the pulmonary artery and aorta).

hepatic portal vein Vein that delivers blood from the stomach, small and large intestines, gallbladder, spleen, and pancreas via capillaries to the liver. It is formed by the joining together of the superior mesenteric vein and the splenic vein and is typically around 3 inches (8 cm) long in adults.

hilum Opening on lung at which the main bronchus (airway) and blood vessels enter; the same term is used for the similar entry point on other organs, such as the spleen.

lungs Twin sacs in the thoracic cavity that together form a key organ in the respiratory system, where in a process called "gaseous exchange" oxygen from inhaled air passes from the lungs into the bloodstream and carbon dioxide passes from the bloodstream into the lungs to be exhaled. The right lung is larger than the left, which has the heart to accommodate on the left side of the chest; they are both protected by the rib cage. Every day a typical person takes around 25,000 breaths and draws in around 2,640 gallons (10,000 liters) of air.

myocardium Type of muscle found in the heart, with the capacity to work without stopping. Myocardium is the middle layer of the heart's muscular wall, lying between the outer epicardium and inner endocardium.

pericardium Double-walled sac that encloses and protects the heart and the roots of the superior vena cava, inferior vena cava, pulmonary artery, and aorta, the four vessels that bring blood to and from the heart.

pleura Double-layered membrane that covers and protects each lung. One part (the parietal pleura) attaches to the wall of the thoracic cavity, while the other (the visceral pleura) is connected to the surface of the lung. The two pleurae touch one another for only a short space to the rear of the breastbone.

pulmonary circulation The part of the body's circulatory system that carries oxygenated blood from the lungs to the heart and deoxygenated blood back to the lungs.

septum The central partition that divides the chambers of the heart. (The same word is used to describe a separating wall in other body parts—for example, the area of cartilage and bone that divides the nostrils in the nose.)

sinoatrial node Also known as the sinus node, a group of specialized cells in the wall of the heart's right atrium that release electrical impulses to stimulate and regulate the contractions of the cardiac cycle. It is often called the heart's natural pacemaker.

systemic circulation The part of the circulatory system that carries blood rich in oxygen from the heart to the organs and tissues of the body and returns the deoxygenated blood to the heart. The lungs and heart have their own systems (respectively known as the pulmonary and the coronary circulation).

trachea Also known as windpipe, the flexible fibrocartilaginous tube (one containing fibrous tissue and cartilage) that leads from the larynx to the lungs and branches to form the right and left bronchi. Each of the bronchi branches up to 25 times into smaller and smaller airways, terminating at the tiny sacs called pulmonary alveoli, where gaseous exchange takes place through the thin membranous walls of the alveoli.

ventricle One of the two lower chambers of the heart.

THE CIRCULATORY SYSTEM

the 30-second anatomy

RELATED TOPICS

See also
THE HEART
page 64
THE MAJOR ARTERIES & VEINS
page 66
THE MICROCIRCULATION
page 68

3-SECOND INCISION

The circulatory system is a series of tubes or vessels through which blood is transported around the human body.

3-MINUTE DISSECTION

Blood normally flows smoothly, which stimulates cells lining arteries to release chemicals that protect the vessels and keep them open. However, where an artery divides into two, the blood flow becomes chaotic and these protective chemicals are lost. This makes it easier for fatty deposits from the blood to enter the artery wall, build up, and eventually cause a blockage. If this process occurs in an artery supplying heart muscle, it can cause a heart attack.

Blood is transported around the body to deliver nutrients and oxygen to tissues and organs, and to remove unwanted chemicals and carbon dioxide to parts of the body where these can be destroyed or expelled. Blood is carried toward the heart in veins and away from the heart in arteries. The circulatory system can be broadly divided into two parts. The pulmonary system consists of vessels carrying blood low in oxygen from the right side of the heart to the lungs, where the oxygen content is replenished before the blood travels back to the left side of the heart. The systemic circulation then transports this blood to the rest of the body to deliver oxygen and nutrients to tissues, before returning it to the right side of the heart. The whole process repeats as the blood is circulated around the body. The beating of the heart provides the driving force that maintains the movement of blood around the circulation.

3-SECOND BIOGRAPHY
WILLIAM HARVEY
1578–1657
An English physician who discovered the circulatory system

30-SECOND TEXT
Andrew T. Chaytor

Blood in arteries running to the body is usually bright red (because charged with oxygen), while blood in veins (here shown as blue) is a dull red.

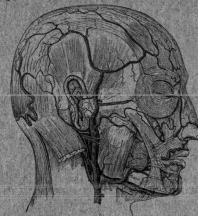

MAJOR ARTERIES OF
THE FACE & SCALP

MAJOR VEINS OF THE
HEAD & NECK

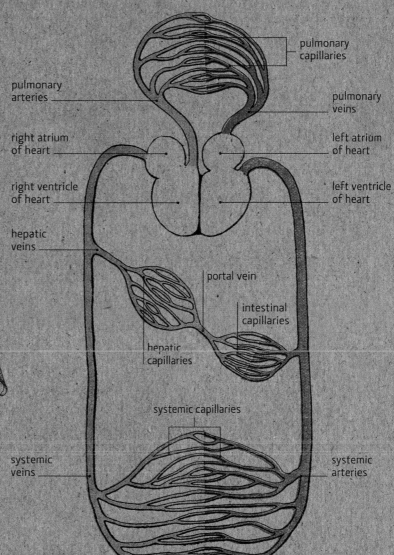

pulmonary
capillaries

pulmonary
arteries

pulmonary
veins

right atrium
of heart

left atrium
of heart

right ventricle
of heart

left ventricle
of heart

hepatic
veins

portal vein

intestinal
capillaries

hepatic
capillaries

systemic capillaries

systemic
veins

systemic
arteries

GENERAL CIRCULATION
OF THE BLOOD

THE HEART

the 30-second anatomy

A muscle about the size of a clenched fist, the human heart sits toward the front of the chest on the left side. The heart muscle contains within it two upper spaces or chambers (the left and right atria) and two lower chambers (the left and right ventricles), which are filled with blood. The left atrium and left ventricle are connected, as are the right atrium and right ventricle—but there is no connection between the chambers of the left and right sides of the heart. Blood enters the heart via veins that drain into the atria. The blood then moves into the ventricles and exits the heart via the arteries. The role of the heart is to squeeze the blood in the two ventricles sufficiently for it to rush out of the heart and into the arteries for transport around the body. Blood moves in only one direction through the heart—from the veins to the atria to the ventricles to the arteries. This is because of the action of one-way valves that sit where the atria connect to the ventricles, and also where the ventricles connect to the arteries.

3-SECOND INCISION
The heart is a muscular organ that receives blood from veins and then contracts, squeezing blood under pressure into arteries to be transported around the body.

3-MINUTE DISSECTION
The heartbeat is the sound of heart valves closing. The first part ("lub") is the closing of the mitral valve (between left atrium and ventricle) and tricuspid valve (between right atrium and ventricle); the second ("dup"), the shutting of the aortic valve (between left ventricle and aorta) and pulmonary valve (between right ventricle and pulmonary artery). On average, the human heart beats 72 times per minute—equivalent to 4,320 times per hour or 37,843,200 times per year.

RELATED TOPICS
See also
THE CIRCULATORY SYSTEM
page 62
THE MAJOR ARTERIES & VEINS
page 66
THE AUTONOMIC NERVOUS
SYSTEM
page 134

3-SECOND BIOGRAPHY
LUDWIG REHN
1849–1930
A German surgeon who performed the first successful heart surgery in 1896

30-SECOND TEXT
Andrew T. Chaytor

Muscle in the left side of the heart, which pumps blood to the body, is between three and six times thicker than muscle on the heart's right side.

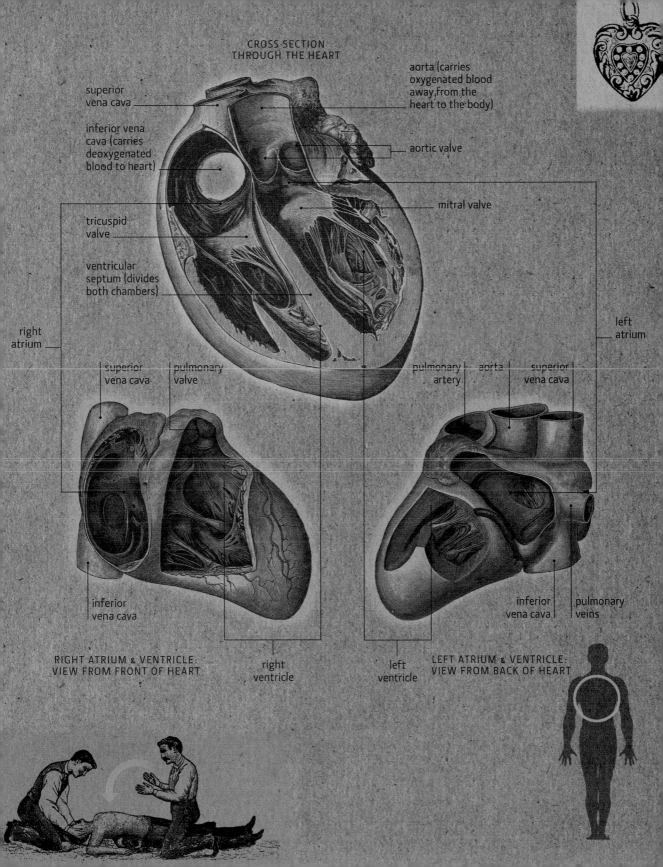

CROSS SECTION
THROUGH THE HEART

superior
vena cava

inferior vena
cava (carries
deoxygenated
blood to heart)

tricuspid
valve

ventricular
septum (divides
both chambers)

right
atrium

aorta (carries
oxygenated blood
away from the
heart to the body)

aortic valve

mitral valve

left
atrium

superior
vena cava

pulmonary
valve

pulmonary
artery

aorta

superior
vena cava

inferior
vena cava

inferior
vena cava

pulmonary
veins

RIGHT ATRIUM & VENTRICLE:
VIEW FROM FRONT OF HEART

right
ventricle

left
ventricle

LEFT ATRIUM & VENTRICLE:
VIEW FROM BACK OF HEART

THE MAJOR ARTERIES & VEINS

the 30-second anatomy

The major arteries of the human body all carry oxygenated blood away from the heart, with one exception—the pulmonary artery, which carries blood low in oxygen from the right side of the heart to the lungs. The body's largest artery is the aorta, which carries blood that has been pumped out from the left side of the heart under high pressure. The aorta has a thick muscular wall that also contains an elastic material; the stretchiness of the wall keeps the blood under high pressure and helps to push the blood to the more distant parts of the body. The aorta divides into smaller arteries, which have smaller amounts of elastic material in their walls but a greater proportion of muscle. Once blood flows through a tissue it is under low pressure and enters the small veins. The small veins lead to larger veins, which have thin, stretchy walls, containing smaller amounts of muscle. The superior and inferior vena cava are veins that return blood low in oxygen back to the right side of the heart; the pulmonary vein is the only vein to carry oxygenated blood—it runs from the lungs to the left side of the heart.

3-SECOND INCISION
The major arteries are vessels that carry blood away from the heart, while the major veins carry blood toward the heart.

3-MINUTE DISSECTION
Veins hold approximately two-thirds of all the blood in the body—acting like a reservoir for blood. The stretchiness of their walls makes this possible. Veins are generally less muscular than arteries and are usually closer to the skin. Some arteries are close enough to the skin for a person to feel the pulse of blood; the best are the radial artery at the wrist and the brachial artery on the inside of the elbow.

RELATED TOPICS
See also
THE CIRCULATORY SYSTEM
page 62
THE HEART
page 64
THE MICROCIRCULATION
page 68

3-SECOND BIOGRAPHY
PRAXAGORAS OF COS
Active ca. 340 BCE
An influential early Greek medical figure who is credited with being the first to distinguish between arteries and veins

30-SECOND TEXT
Andrew T. Chaytor

The carotid arteries carry oxygenated blood to the head, while the jugular veins return deoxygenated blood to the heart.

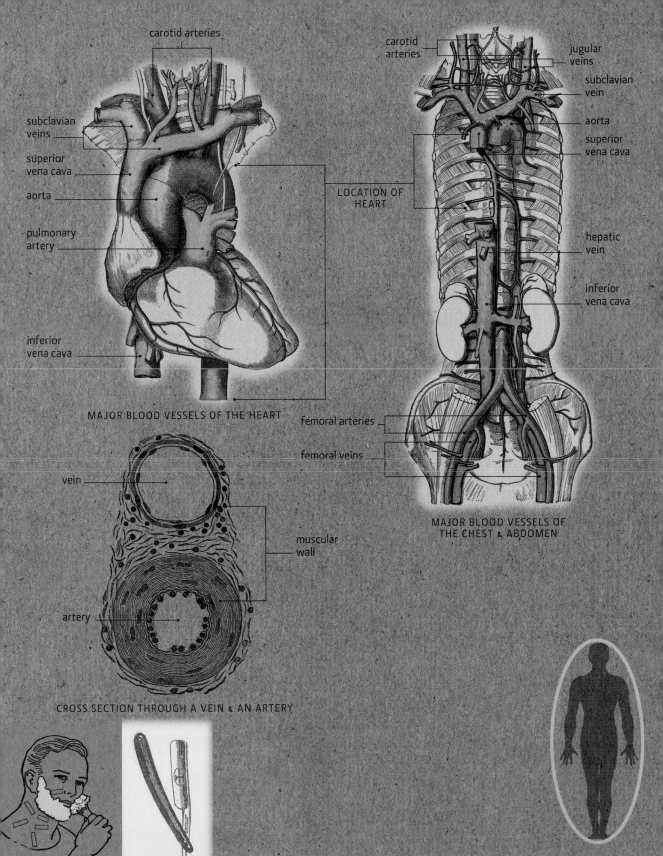

carotid arteries

subclavian
veins

superior
vena cava

aorta

pulmonary
artery

inferior
vena cava

MAJOR BLOOD VESSELS OF THE HEART

carotid
arteries

jugular
veins

subclavian
vein

aorta

superior
vena cava

LOCATION OF
HEART

hepatic
vein

inferior
vena cava

femoral arteries

femoral veins

MAJOR BLOOD VESSELS OF
THE CHEST & ABDOMEN

vein

muscular
wall

artery

CROSS SECTION THROUGH A VEIN & AN ARTERY

THE MICROCIRCULATION

the 30-second anatomy

3-SECOND INCISION
The microcirculation—a collection of the smallest blood vessels in the body—controls blood flow through an organ.

3-MINUTE DISSECTION
When fluid movement across the microcirculation goes wrong it can lead to tissue dehydration or swelling (edema). Elephantiasis is a parasitic worm infection transmitted by mosquitos; the parasite blocks the lymphatic drainage system, leading to extreme swelling— particularly in the lower body. In contrast, during hemorrhage (when blood discharges from vessels), up to 1 pint (500 ml) of tissue fluid can move into the capillaries to boost blood volume in a process called autotransfusion.

The microcirculation consists of the arterioles, capillaries, and venules, which together supply blood to (and drain blood from) the organs. The arterioles are approximately 5–100 microns in diameter and have an outer layer of thick, smooth muscle, a thinner adventitial (membranous) layer, and an inner lining of thin endothelial cells. (1 micron, also called 1 micrometer, is 1/1,000,000 meter.) The extent to which the smooth muscle layer is contracted (dilation or constriction) determines the diameter of the arteriole and, hence, the flow of blood through the vessel and into the capillaries. In diameter the capillaries are 5–10 microns and consist of only the thin endothelial cell layer. Some capillaries have a diameter smaller than a red blood cell, requiring blood cells to deform to flow through these tiny vessels. The structure of capillaries is well suited to their function. They provide a thin barrier across which oxygen and nutrients, such as glucose, can cross and carbon dioxide can leave. This blood flow is called "nutritive flow." Venules are primarily collecting channels that drain into the larger veins. Fluid is filtered across the walls of the capillaries and venules to regulate levels of water (hydration) in tissue.

RELATED TOPICS
See also
THE LYMPHATIC SYSTEM
page 100

3-SECOND BIOGRAPHY
MARCELLO MARPIGHI
1628–1694
An Italian physiologist who confirmed the existence of capillary circulation four years after the death of William Harvey

30-SECOND TEXT
Marina Sawdon

In a capillary bed, a network of capillaries supplies oxygenated blood to—and drains deoxygenated blood from—body tissues.

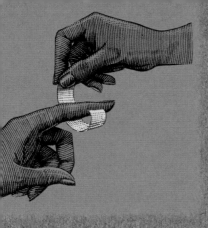

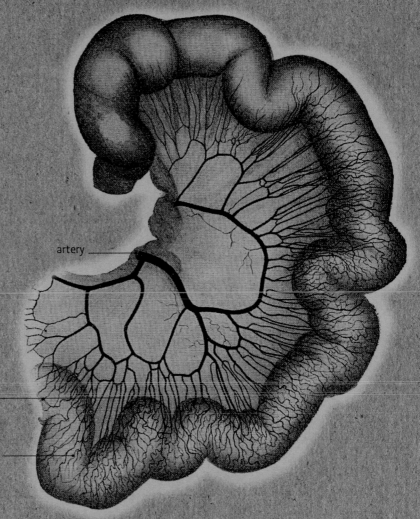

artery

arterioles

capillaries

CAPILLARY NETWORK IN
THE SMALL INTESTINE

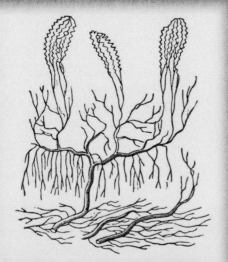

THE PORTAL CIRCULATION

the 30-second anatomy

A portal system consists of a

vein that originates in a grouping of the smallest blood vessels in the body (a capillary bed) and travels directly to a second capillary bed in a different tissue. The largest portal system in the body is the hepatic portal system, which carries blood low in oxygen but high in nutrients from the intestines to the liver via the porta hepatis (gateway of the liver and the origin of the name "portal system"); because the hepatic portal vein receives no real benefit from the beating of the heart, the blood remains under low pressure. A key advantage of the hepatic portal system is that useful products of digestion can be transported directly to the liver for storage or processing. Other portal systems occur in the brain and kidneys. The hypothalamus in the brain produces releasing hormones that travel via a direct portal system to the pituitary gland, where they control release of other hormones; in the kidney, a portal system connects the parts of the kidney responsible for filtering blood to the part that prevents excessive fluid loss.

3-SECOND INCISION
Portal systems in the human body provide a link between distant organs and are useful for transporting nutrients and signals.

3-MINUTE DISSECTION
Cirrhosis of the liver can be caused by excessive alcohol consumption and viral infections. The liver becomes damaged and does not function properly and scar tissue forms. The scarring interrupts blood flow in the portal vein and causes the blood to back up in the portal system. This forces fluid out of the portal vein and causes a swollen abdomen that may be seen in patients with liver cirrhosis.

RELATED TOPICS
See also
THE MAJOR ARTERIES & VEINS
page 66
THE MICROCIRCULATION
page 68

30-SECOND TEXT
Andrew T. Chaytor

Around 3 inches (8 cm) long, the hepatic portal vein carries three-quarters of the blood in the hepatic portal system, while the rest is carried in the hepatic arteries.

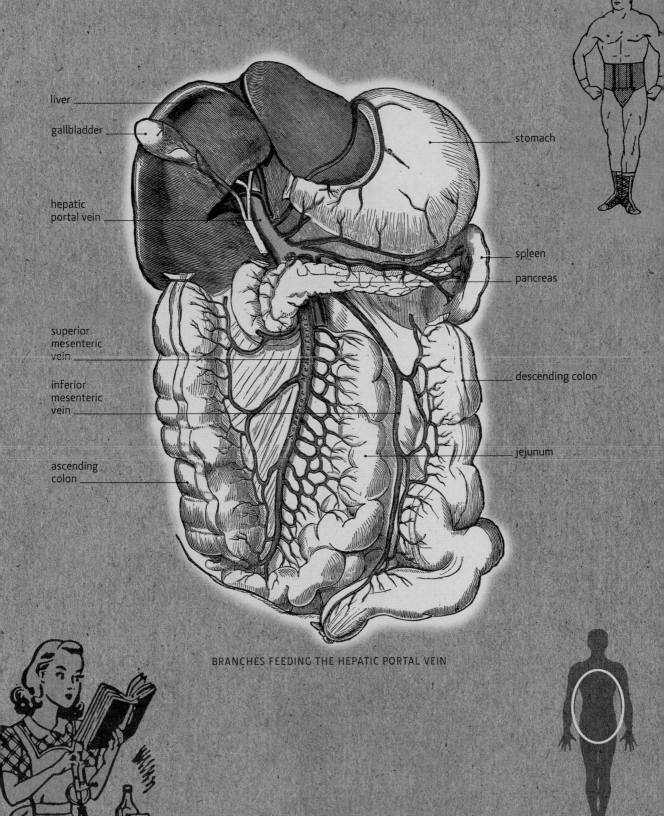

liver

gallbladder

hepatic
portal vein

superior
mesenteric
vein

inferior
mesenteric
vein

ascending
colon

stomach

spleen

pancreas

descending colon

jejunum

BRANCHES FEEDING THE HEPATIC PORTAL VEIN

THE SPLEEN

the 30-second anatomy

RELATED TOPICS
See also
THE LYMPHATIC SYSTEM
page 100

3-SECOND INCISION
The spleen destroys old or damaged red blood cells and filters pathogens circulating in the bloodstream; it is composed of red and white pulp.

3-MINUTE DISSECTION
When the function of the bone marrow is compromised, the spleen reverts to the function it performs in the fetus—the manufacture of red blood cells. Following injury and severe blood loss, the spleen contracts and so forces stored blood into circulation. The spleen can be removed if it becomes diseased or injured, and the body will continue to function. However, people without a spleen are more vulnerable to infections.

The spleen is found between the diaphragm, the stomach, and the left kidney, extending from the 9th and 10th rib in the left upper quadrant of the abdomen. The largest lymphoid organ in the body, it weighs up to $5\frac{1}{3}$ ounces (150 g) and is around $4\frac{1}{2}$–5 inches (12 cm) long—the size of a clenched fist. It is a deep red color due to the unusual amount of blood it contains and performs processes on blood similar to the ones lymph nodes perform on lymph. A network of arterial channels and spaces (sinuses) act like a strainer to slow the blood flow through the spleen, aiding the filtering of the blood, and activate an immune response if required. The splenic hilum is an opening through which the splenic artery enters and the splenic vein and lymphatic draining vessels leave. The spleen's cellular contents are referred to as pulp. Large numbers of red blood cells are found within the red pulp, while within the white pulp are areas of lymphoid nodules that synthesize antibodies. The spleen is vulnerable to rupture during abdominal trauma. A damaged spleen requires immediate treatment because large amounts of blood can leak into the abdominal cavity.

3-SECOND BIOGRAPHY
RUDOLF VIRCHOW
1821–1902
A German pathologist who first described the term leukemia in 1856

30-SECOND TEXT
Jo Bishop

The spleen is supported in its position on the upper left part of the abdomen by two ligaments that connect it to the stomach and the left kidney.

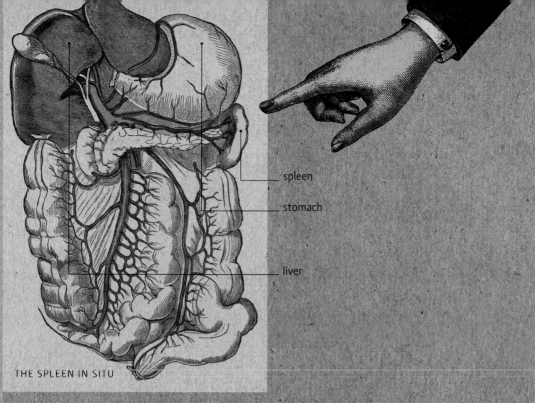

spleen

stomach

liver

THE SPLEEN IN SITU

CROSS SECTION THROUGH THE SPLEEN

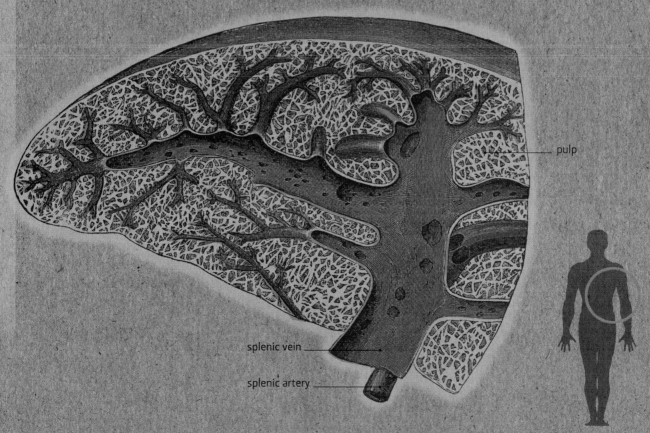

pulp

splenic vein

splenic artery

1578
Born Folkestone, Kent

1597
Graduated from Gonville and Caius College, Cambridge

1604
Joined the College of Physicians, London

1609
Appointed Physician at St. Bartholomew's Hospital, London

1613
Elected Censor of the College of Physicians. The censor's task was to examine applicants to the college and maintain medical standards

1615–1656
Appointed Lumleian Lecturer to the College of Physicians; obliged to deliver public lectures for seven years

1616
In anatomy lectures, gives first description of his ideas on circulation

1618
Appointed Physician Extraordinary to King James I

1625
Reelected Censor at the College of Physicians

1628
Published theories in *Exercitatio Anatomica de Motu Cordis et Sanguinis in Animalibus* ("An Anatomical Study of the Motion of the Heart and of the Blood in Animals")

1632
Appointed Physician in Ordinary to King Charles I

1654
Elected President of the College of Physicians, but could not accept because of ill health

1657
Died at Roehampton, London; buried at Hempstead, Essex

WILLIAM HARVEY

You could describe William

Harvey as the Galileo of medical science. (It's very tempting to think that the two men met, because Harvey was in Rome at the same time as Galileo, in 1636, but there is no supporting evidence.) Harvey gave us the first accurate description of the circulatory system and of the heart's role in this system as a pump, overthrowing the reigning orthodoxy based on Galen's theories (see pages 114—115). Putting the heart at the center of the system had all the compelling elegance of Galileo's heliocentric theory (which put the Sun at the center of the universe), and courted just as much controversy. Harvey did not incur the anger of popes, as Galileo had, but to some extent the medical establishment did close ranks against him, with many doctors asserting that they would "rather err with Galen than proclaim the truth with Harvey." He lost out financially and his practice suffered—but his career was nevertheless successful. An astute marriage in 1604 to Elizabeth Browne (daughter of Lancelot Browne, court physician formerly to Queen Elizabeth I and by this stage to her successor, King James I) meant that by the time Harvey published his ideas, he was court physician to King Charles I, with a significant proportion of the aristocracy on his books; he also held prestigious posts at the College of Physicians; and more or less ran St. Bartholomew's Hospital, set up pro bono to help the poor. Consequently, his revolutionary stance did not cast him into darkness.

Harvey's ideas were sparked by his mentor at the University of Padua, Hieronymous Fabricius, who made a study of the one-way valves that he deduced must orchestrate the venous system. Harvey extrapolated from this, concluded that the venous system returned blood to the heart (rather than dispersing it somehow through the skin), that the lungs must be an oxygenator rather than just a cooling mechanism, and that arterial blood was not, therefore, new, but reoxygenated. Lack of sufficiently powerful microscopy meant that Harvey could not see how the capillaries worked, but he deduced that they must exist (see pages 62–67). He could calculate how fast the heart beat and how much blood is pumped around the system.

Ironically, Harvey died of a cerebral hemorrhage brought on by gout. There is a research institute named after him at St. Bartholomew's Hospital and an eponymous modern hospital in Ashford, Kent, England.

THE LUNGS

the 30-second anatomy

At tiny sacs called alveoli in the lungs, a process called "gaseous exchange" allows for inhaled oxygen to permeate into the blood to be carried to the heart, and carbon dioxide in blood arriving from the heart to pass into the lungs to be exhaled. The two lungs are set with one on each side of the heart, within the thoracic cavity, with their base on the diaphragm, a pointed top (apex) projecting into the neck, a rib (costal), and middle surface. The lung tissue (parenchyma) contains elastic fibers, smooth muscle, and lymphatics; it is divided into lobules supplied by tributaries from pulmonary arteries and veins and respiratory passageways. Structures enter and leave the lung at its hilum (doorway). These include the pulmonary artery that brings deoxygenated blood from the heart, two pulmonary veins that carry oxygenated blood to the heart, and the main bronchus, bronchial vessels, nerves, and lymphatics. The right lung has three lobes—superior, middle, and inferior—that are separated by an oblique fissure. The horizontal fissure separates the superior lobe from the middle lobe. The left lung is smaller; its two lobes—superior and inferior— are separated by an oblique fissure.

3-SECOND INCISION
In the lungs, oxygen permeates from inhaled air into the blood, and carbon dioxide passes from the blood into the lungs to be exhaled.

3-MINUTE DISSECTION
Each lung is covered by a continuous double-layered sac—the pleura. Parietal pleura is attached to the thoracic wall and visceral pleura attaches to the surface of the lungs. The pleural space between them is filled with a lubricant, surfactant. Premature babies struggle to breathe because they do not produce surfactant. They are given artificial surfactant and ventilated until their lungs mature.

RELATED TOPICS
See also
THE RESPIRATORY MUSCLES
page 56
THE BRONCHIAL TREE
page 78

3-SECOND BIOGRAPHY
KURT VON NEERGAARD
1887–1947
A Danish doctor who in 1929 described the function of surfactant

30-SECOND TEXT
Jo Bishop

The heart sits in the chest cavity between the lungs, to which it is connected by the pulmonary artery and two pulmonary veins.

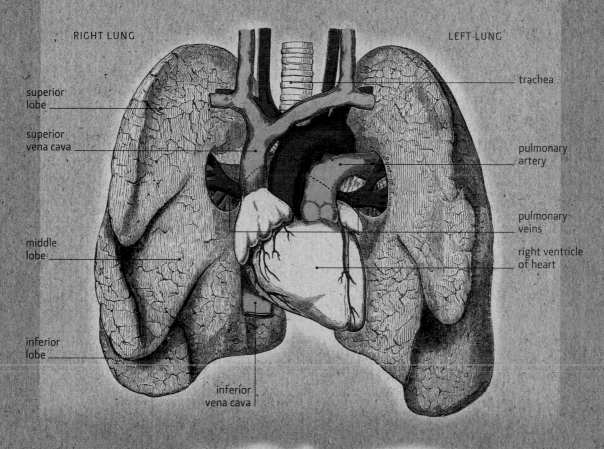

RIGHT LUNG

LEFT LUNG

superior lobe

trachea

superior vena cava

pulmonary artery

pulmonary veins

middle lobe

right ventricle of heart

inferior lobe

inferior vena cava

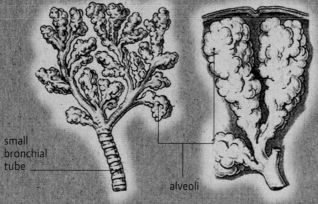

small bronchial tube

alveoli

BRONCHIAL TUBE & ALVEOLI

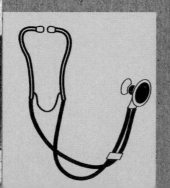

THE BRONCHIAL TREE

the 30-second anatomy

3-SECOND INCISION
The bronchial tree conducts air to the lungs—it consists of the trachea, right and left main bronchi, lobar bronchi, segmental bronchi, and terminal bronchioles.

3-MINUTE DISSECTION
There are approximately 300 million alveoli. Their surface area is around 645 square feet (60 m²). The average person breathes about 25,000 times a day; the lung is very light—if a piece of lung is placed in water, it will float.

The trachea, or windpipe, conducts air from the larynx to the lungs. It is a flexible tube—on average about 4⅓ inches (11 cm) long and 1 inch (2.5 cm) in diameter—kept open and supported by a series of C-shaped cartilages; the incomplete ring lets food pass easily into the esophagus behind. The trachea branches to become the right and left main bronchi, which also contain supporting cartilage; they lead respectively to the right and left lung. Further branching occurs within the lung tissue. The primary bronchus divides to form lobar bronchi (also known as secondary bronchi), three on the right and two on the left, which then branch to form segmental (tertiary) bronchi. Tertiary bronchi supply air to specific regions of the lung—the bronchopulmonary segment. This branching continues through the lung tissue, with the passages at each stage becoming progressively smaller, as in the branches of a tree—hence the name "bronchial tree." The amount of supporting cartilage reduces, until the smallest conducting bronchioles contain only smooth muscle. Gaseous exchange takes place at the terminal bronchiole within clusters of sacs called alveoli. Here, inhaled oxygen enters the bloodstream and carbon dioxide leaves it. The carbon dioxide is then expelled from the body when exhaling.

RELATED TOPICS
See also
THE RESPIRATORY MUSCLES
page 56
THE LUNGS
page 76
THE PHARYNX, LARYNX &
VOCAL CORDS
page 120

3-SECOND BIOGRAPHY
ROBERT BOYLE
1627–1691
An Anglo-Irish chemist who devised Boyle's law for calculating gas exchange

30-SECOND TEXT
Jo Bishop

The right main bronchus is wider and descends at a steeper angle than the left; both bronchi continue branching down to the terminal bronchioles.

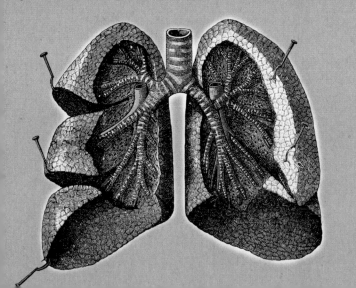

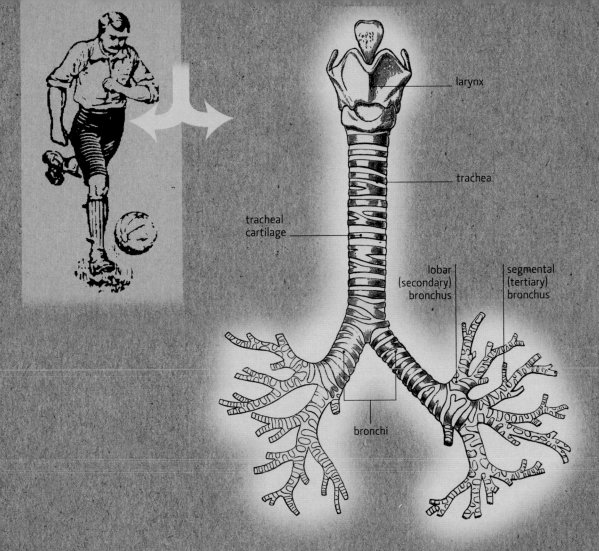

larynx

trachea

tracheal
cartilage

lobar
(secondary)
bronchus

segmental
(tertiary)
bronchus

bronchi

THE BRONCHIAL TREE IN SITU

THE DIGESTIVE SYSTEM

ampulla of Vater Dilated part of the bile duct that meets with the pancreatic duct, then opens into the duodenum, the first part of the small intestine.

bile duct Passageway running from the gallbladder to the duodenum. The bile duct carries bile, a bitter, green-brown liquid produced by the liver and stored in the gallbladder; bile plays a key role in the digestion of food, especially the breakdown of fats, in the small intestine.

esophagus The first section of the gastrointestinal tract, a muscular tube that runs from the throat to the stomach. Its name derives from the ancient Greek *oisophagos* ("eating entrance").

gastrointestinal tract Also known as alimentary canal, the tube running from the mouth to the anus in which food is digested by muscular movement and the action of hormones and enzymes. The tract passes from the mouth through the esophagus, stomach, small intestine, large intestine, and rectum to the anus. The wavelike sequence of muscular contractions that move the mixture along the tube is called peristalsis.

ileocecal junction In the right side of pelvis, the point at which the ileum, the last part of the small intestine, joins the cecum at the start of the large intestine. There are five sections to the large intestine—the cecum, ascending colon, transverse colon, descending colon, and sigmoid colon—throughout which the body continues digestion of the food mixture and absorbs water.

islets of Langerhans Specialized cells in the pancreas that produce the hormones insulin and glucagon (which respectively lower and raise blood glucose levels), and somatostatin (which controls release of growth hormone).

lymph nodes Also known as lymph glands, bean-shaped swellings found at intervals in the system of lymphatic vessels. Lymph nodes contain white blood cells that serve to attack bacteria and combat infection.

pancreatic duct Passageway leading from the pancreas that meets with the bile duct at the ampulla of Vater prior to opening into the duodenum. The duct carries enzymes that aid digestion of food.

pylorus Third of three parts of the stomach, after the fundus and the stomach body. From the pylorus the partly digested liquid food mixture (chyme) passes through the pyloric sphincter into the duodenum, the first section of the small intestine; the sphincter will close to prevent any remaining food particles in the liquid from passing into the small intestine. Food passes into the small intestine about one hour after eating a light meal and up to seven hours after a heavy meal.

rectum The final section of the large intestine, which runs downward through the pelvis. The rectum is around 9 inches (23 cm) long; it passes through the anal canal to the external opening of the anus.

renal cortex Region of the kidney that filters blood and makes urine; the urine filtering through the renal medulla before flowing into the urinary bladder by way of the ureter. The cortex contains 1.2 million tubules called nephrons; as well as filtering blood and forming urine, they reabsorb a number of useful substances. A nephron consists of a collection of blood capillaries called a glomerus within a sac (Bowman's capsule). There are two kidneys, one on each side of the spine; the kidneys filter $2\frac{3}{4}$ pints (1.3 liters) of blood per minute.

renal medulla Inner portion of the kidney, containing the renal pyramids, minor calyx, pelvic space, and major calyx.

sinusoids Tiny spaces in the liver lined with hepatocyte cells that detoxify the blood and produce bile. The bile is stored in the gallbladder on the underside of the liver, and flows from there along the bile duct into the small intestine.

ureters Two tubes that drain urine from the kidneys to the urinary bladder. There is one ureter for each kidney. The ureters are about 12 inches (30 cm) long.

villi Tiny projections on the inner walls of the jejunum and ileum in the small intestine. Through the villi nutrients are absorbed from the food mixture in the process of digestion.

THE STOMACH
the 30-second anatomy

The stomach is the second part
of the gastrointestinal tract after the esophagus, the tube that brings chewed food from the mouth to the stomach. It is in the upper left part of the abdomen, set apart from the heart and the left lung by the diaphragm. The walls of the stomach contain three muscle layers that enable the stomach to contract in different ways to make sure that food is mixed with gastric juices. The gastric juices are produced by cells lining the stomach and include acidic substances and enzymes to break up food. The stomach is divided into three parts: fundus, body, and pylorus. The fundus is the stomach's dilated upper portion and often contains gas, although it may also contain fluid and/or food. In the body of the stomach, the inner walls, which are formed into folds known as rugae, contain gastric glands to aid digestion. Once food has been broken down, it passes into the pylorus, where a band of muscle (sphincter) acts like a door and lets the food pass into the small intestine.

3-SECOND INCISION
The stomach is a dilated part of the gastrointestinal tract, and is where the body begins the process of digesting food.

3-MINUTE DISSECTION
Acid levels inside the stomach can be 1.5pH— more than vinegar or lemon juice but not quite as much as battery acid! Army surgeon William Beaumont (1785–1853) experimented with a patient who had survived a gunshot in the stomach, putting food into the stomach via the entry wound. This led to the discovery of stomach acid, and the understanding that digestion is largely a chemical and not a mechanical process.

RELATED TOPICS
See also
THE SMALL INTESTINE
page 86
THE LARGE INTESTINE
page 88

3-SECOND BIOGRAPHY
WILLIAM BEAUMONT
1785–1853
A researcher into human digestion and the discoverer of stomach acid

30-SECOND TEXT
Claire France Smith

The stomach's size and shape vary according to an individual's posture and physique; in an adult its average capacity is just over 3 pints (1.5 liters).

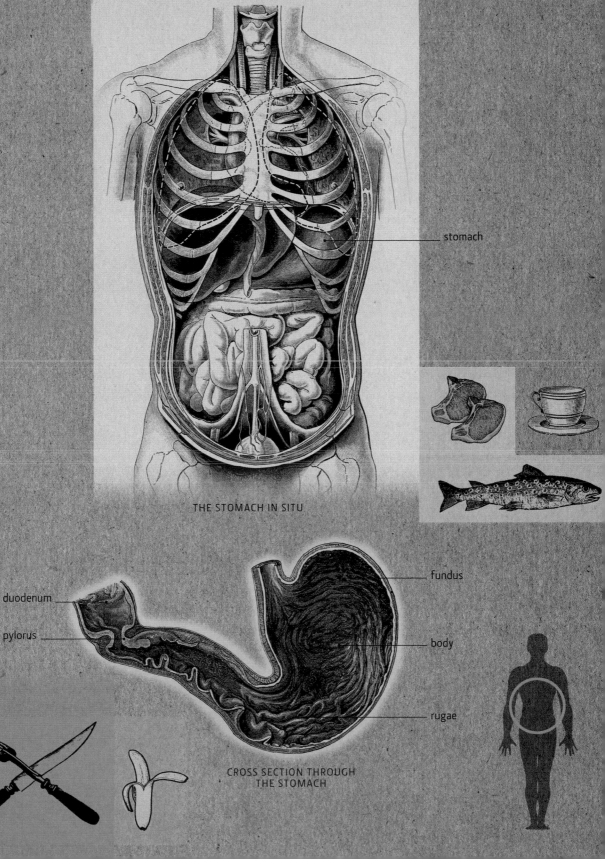

stomach

THE STOMACH IN SITU

duodenum

pylorus

fundus

body

rugae

CROSS SECTION THROUGH
THE STOMACH

THE SMALL INTESTINE

the 30-second anatomy

The gastrointestinal tract continues from the stomach into the small intestine, which is divided into three parts: the duodenum, jejunum, and ileum. The duodenum receives secretions from the bile duct and pancreatic duct that contain enzymes to aid digestion. The jejunum and ileum both have the same function—to absorb nutrients from the food. To increase their surface area, and so improve the absorption of nutrients, the inside walls contain small folds that hold millions of tiny projections (villi). There is no line of demarcation between the jejunum and ileum, although the ileum tends to be of a smaller diameter and contains areas of lymph known as Payer's patches, which help protect the body by providing an immune response to any potentially harmful element from the external environment. The small intestine is very long, approximately 23 feet (7 m), and folds itself to fit within the abdomen. It is free to move about to a certain extent, but to prevent it from becoming twisted, it is anchored to the back wall of the abdomen by a membrane called the mesentery. The last part of the small intestine (ileum) ends at the ileocecal junction, which is in the lower right area of the pelvis.

3-SECOND INCISION
The small intestine is the part of the gastrointestinal tract after the stomach, and is where digestion of food continues and nutrients are absorbed.

3-MINUTE DISSECTION
The wall lining the duodenum contains Brunner's glands, named after Johann Conrad Brunner (1653–1727), who first described them. These glands secrete an alkaline substance to prevent the acidic food contents from burning the walls of the small intestine. The layers of circular muscle that surround the digestive tube contract rhythmically to push the food mixture (chyme) along the tract. This means a person can still digest food when he or she is upside down.

RELATED TOPICS
See also
THE STOMACH
page 84
THE LARGE INTESTINE
page 88

3-SECOND BIOGRAPHY
JOHANN CONRAD BRUNNER
1653–1727
The discoverer of Brunner's glands in the duodenum

30-SECOND TEXT
Claire France Smith

Microscopically small projections (villi) increase the surface area within the small intestine to improve absorption of nutrients.

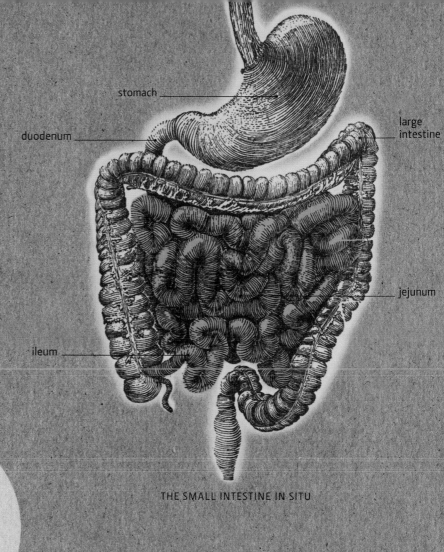

stomach

duodenum

large intestine

jejunum

ileum

THE SMALL INTESTINE IN SITU

VILLI IN LINING OF THE
SMALL INTESTINE

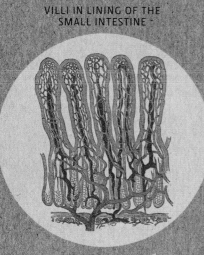

THE LARGE INTESTINE

the 30-second anatomy

The large intestine begins at the ileocecal junction in the right side of the pelvis. There are five parts to the large intestine: cecum, ascending colon, transverse colon, descending colon, and sigmoid colon. The cecum is a pouch that receives the small intestine at the ileocecal valve and forms into the ascending colon. The ascending, transverse, and descending colons go up, across the body and down; the sigmoid colon makes a loop and is continuous with the rectum. To propel the food mixture (chyme) along, the large intestine has a muscular wall with two layers—the outer layer arranged longitudinally and the inner layer in circular bands. This contributes to the large intestine's bumpy appearance. On the outside of the large intestine are fatty tags (appendices epiploicae). Their function is not known; they may act to cushion the colon and possibly are part of the body's immune system. All parts of the large intestine have the same function: to continue digestion and to absorb water. In contrast to the small intestine, the large intestine does not have villi, but it does contain tubular glands (glands of Lieberkühn) that secrete enzymes; other cells secrete mucus to aid the movement of chyme.

3-SECOND INCISION
In the large intestine— the section of the gastrointestinal tract after the small intestine—the food mixture is further digested and its water content absorbed.

3-MINUTE DISSECTION
The appendix, located just off the cecum, is a blind-ended sac with no function so it can be surgically removed without lasting effects if it becomes inflamed (in appendicitis). In humans the cecum is about 4 inches (10 cm) in length, but in animals the cecum/ appendix can be very long; for example, in horses it can measure up to 3 feet 4 inches (1 m) and contains enzymes for processing grass-based food.

RELATED TOPICS
See also
THE STOMACH
page 84
THE SMALL INTESTINE
page 86

3-SECOND BIOGRAPHY
JOHANN NATHANAEL LIEBERKÜHN
1711–1756
A German anatomist who first described intestinal glands in the large intestine

30-SECOND TEXT
Claire France Smith

The large intestine is much shorter than the small intestine—5 feet (1.5 m) compared to 23 feet (7 m); it is called large because it is about twice as wide.

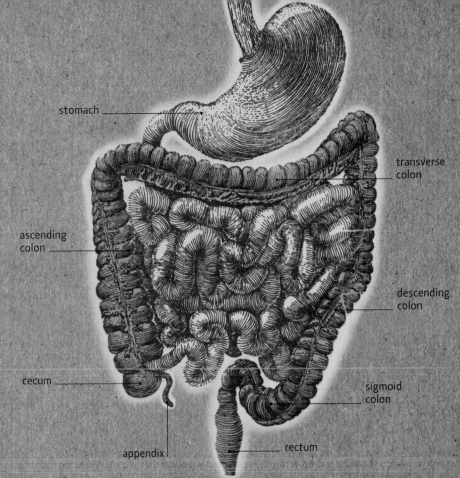

stomach

transverse
colon

ascending
colon

descending
colon

cecum

sigmoid
colon

appendix

rectum

THE LARGE INTESTINE IN SITU

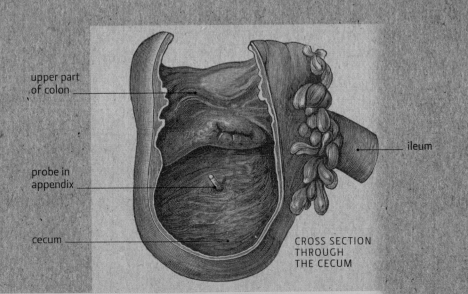

upper part
of colon

ileum

probe in
appendix

cecum

CROSS SECTION
THROUGH
THE CECUM

THE LIVER & GALLBLADDER

the 30-second anatomy

3-SECOND INCISION

The liver is the body's largest gland; it removes toxicants, produces bile to assist digestion, and breaks down red blood cells. The gallbladder stores bile for use by the small intestine.

3-MINUTE DISSECTION

As the main site for detoxification of substances from the gastrointestinal tract, the liver is vulnerable to damage. If damage occurs, remaining liver cells can divide to create new cells—replacing lost tissue and restoring the liver's functionality. The gallbladder not only stores bile but also absorbs water from it and, as a result, the bile becomes more concentrated and more powerful at breaking down fats.

The liver is located in the right upper quadrant of the abdomen, just below the diaphragm. It is anatomically split into unequal right and left lobes by the falciform ligament, and functionally divided into eight segments by the distribution of blood vessels. In addition to removing toxicants, producing bile, and breaking down red blood cells, the liver synthesizes and stores glycogen (from carbohydrates) and produces hormones. It receives 80 percent deoxygenated blood from the gastrointestinal tract and spleen via the portal vein and 20 percent oxygenated blood via the hepatic artery. Within the liver lobes, deoxygenated blood flows through spaces called sinusoids. Hepatocyte cells lining the sinusoids detoxify the blood, which flows back to the heart through the inferior vena cava. The hepatocytes also produce bile, which is transported through ducts to the gallbladder. The gallbladder is a pear-shaped sac on the underside of the liver. Its function is to collect and store bile, a dark green/yellow fluid that assists with the breakdown of fat. Food entering the stomach stimulates the production of the chemical cholecystokinin, which causes the gallbladder to release bile into the small intestine through a sphincter known as the ampulla of Vater.

RELATED TOPICS

See also
THE SMALL INTESTINE
page 86

THE PANCREAS
page 92

3-SECOND BIOGRAPHY

ABRAHAM VATER
1684–1751
The discoverer in 1720 of the ampulla of Vater

30-SECOND TEXT

Claire France Smith

Bile produced by the gallbladder may crystallize and cause inflammation or blockages that require the organ's removal in a cholecystectomy.

LIVER

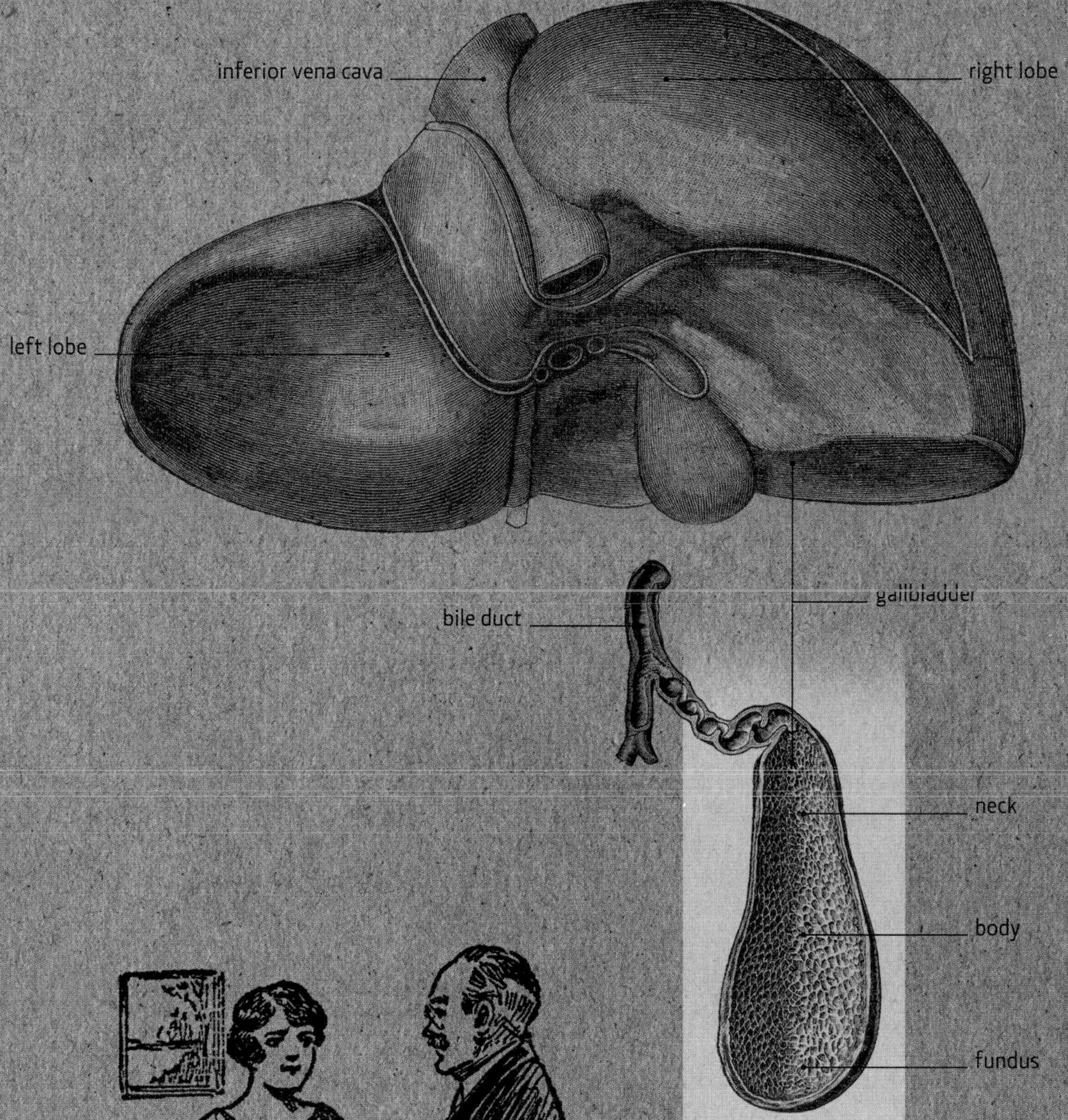

inferior vena cava

right lobe

left lobe

bile duct

gallbladder

neck

body

fundus

**GALLBLADDER:
INTERIOR VIEW**

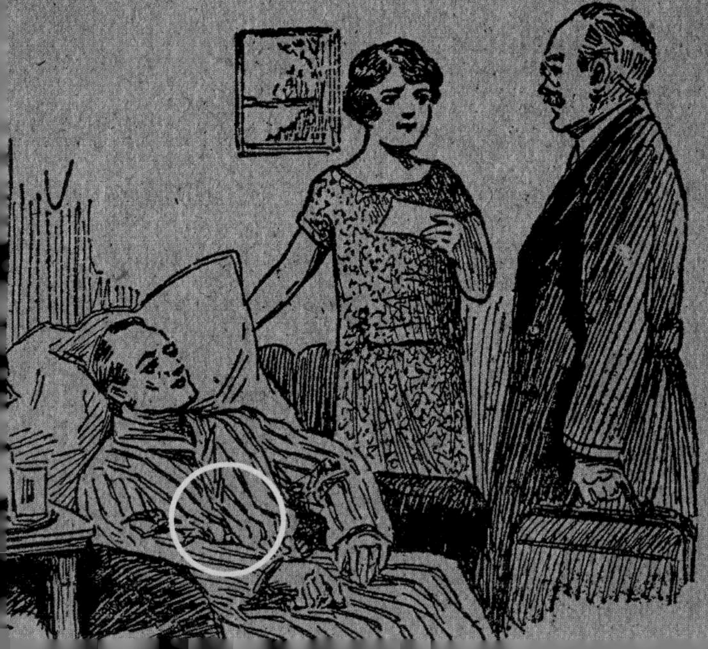

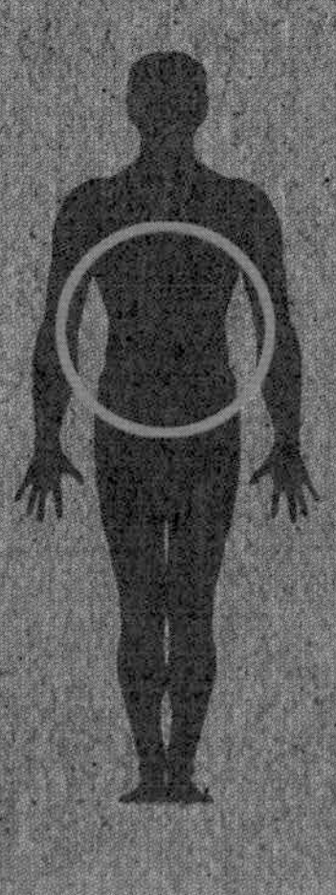

THE PANCREAS

the 30-second anatomy

3-SECOND INCISION
Located behind the stomach, the pancreas secretes enzymes that assist in the digestion of food and hormones that control blood sugar levels.

3-MINUTE DISSECTION
In about 25 percent of people, the main pancreatic duct does not join the bile duct and opens into the duodenum separately: individuals will not know about this because it rarely causes any problems. Diabetes is a condition in which the endocrine cells of the pancreas either do not produce enough insulin or do not respond to the insulin that is produced.

The pancreas has a head, an uncinate process (a U-shaped protuberance), a neck, and a tail. Its head is in the midline of the body and its tail extends to the spleen, under the lower left ribs. The head nestles into the duodenum, the C-shaped first section of the small intestine. The pancreas is a lobulated gland (one with many lobules or small sections of tissue) and its cells are divided into two functional groups: exocrine and endocrine cells. The exocrine cells secrete enzymes that aid food digestion. The enzymes reach the small intestine via ducts; small ducts connect to form the main pancreatic duct, which joins the bile duct to open into the duodenum via the major duodenal papilla. The endocrine cells secrete hormones directly into the bloodstream to control blood sugar levels. Endocrine cells are clustered in regions known as islets of Langerhans, and within these two main cell types can be identified: alpha and beta. Alpha cells secrete glucagon, which raises blood glucose levels, while beta cells have the opposite effect by secreting insulin. To enable the endocrine function and hormones to be distributed by the blood, the pancreas is supplied and drained by numerous pancreatic arteries and veins.

RELATED TOPICS
See also
THE STOMACH
page 84
THE LIVER & GALLBLADDER
page 90
THE ENDOCRINE SYSTEM
page 126

3-SECOND BIOGRAPHY
PAUL LANGERHANS
1847–1888
A German pathologist who discovered the clear clusters of endocrine cells in the pancreas now known as the islets of Langerhans

30-SECOND TEXT
Claire France Smith

The pancreatic duct joins with the bile duct, which brings bile from the liver, before they together open into duodenum.

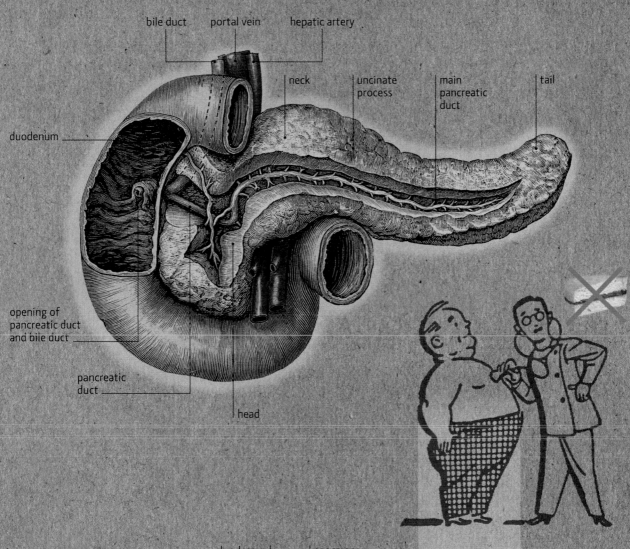

bile duct portal vein hepatic artery

neck uncinate process main pancreatic duct tail

duodenum

opening of pancreatic duct and bile duct

pancreatic duct

head

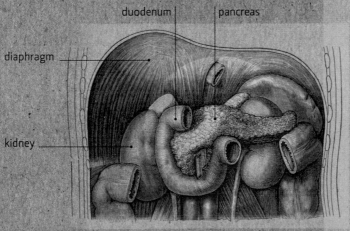

duodenum pancreas

diaphragm

kidney

THE PANCREAS IN SITU

the Digestive System

EUSTACHIUS

A contemporary of Vesalius

(see pages 26–27), Eustachius was a far more circumspect personality; his fear of reprisals and excommunication led him to suppress what would turn out to be his greatest work, and the world did not get to know about it until over a century after his death.

We know Eustachius's name because he gave it to the Eustachian tube, or tuba auditiva, which links the nasopharynx to the middle ear, as well as to the Eustachian valve in the heart's right ventricle, but aside from this not a great deal is known about Bartolomeo Eustachi (Latinized to Eustachius); his birth date and place are obscure, and although it is known he studied medicine in Rome, exactly when is uncertain. We do know that his father, Mariano, was a physician, and that Bartolomeo received an excellent education, learning Greek, Hebrew, and Arabic and making his own translations of the great Persian philosopher and medical author ibn Sina, or Avicenna (ca. 980–1037). Bartolomeo became personal physician to the Duke of Urbino and his brother, Cardinal Giulio della Rovere; thanks to this connection, he became Professor of Anatomy at the Collegia di Sapienza in Rome with access to corpses from the local hospitals.

In addition to his work on the ear, Eustachius was the first anatomist to dedicate a treatise to the kidney, accurately describing the adrenal glands. He is acknowledged to be the first comparative anatomist, because he used animal cadavers to compare with humans and described developmental anatomy, such as infant dentition. He published all his findings, together with some rational defenses of Galen, in *Opuscula Anatomica* (1565).

To illustrate his treatises, and for future such works, Eustachius worked with the artist Pier Matteo Pini to create 47 anatomical plates. The first eight appeared in the *Opuscula*, but thereafter Eustachius suffered a loss of nerve. Fear of the wrath of the Catholic Church prompted him to suppress the other 39 illustrations. The plates were kept in the artist's family, but were not found again until 1714, when they were published, with notes, by Giovanni Maria Lancisi, physician to Pope Clement XI. Artistically, they are less polished than Vesalius's illustrations, but they are extremely accurate and their disappearance meant not only that Eustachius was denied recognition as a founding father of anatomy, alongside Vesalius, but also that anatomical knowledge was held back for a whole century.

THE KIDNEYS

the 30-second anatomy

3-SECOND INCISION
The kidneys—a pair of bean-shaped structures within the abdomen, situated one on each side of the spine—regulate many body functions and also produce urine.

3-MINUTE DISSECTION
The kidneys filter almost 48 gallons (180 liters) of plasma (the pale yellow fluid constituent of blood) daily, yet the body contains only 7½ pints (3.5 liters) of plasma. The kidneys reabsorb most of this liquid under the influence of antidiuretic hormone (ADH). Alcohol inhibits ADH, reducing the amount the kidneys reabsorb. For this reason, a person needs to urinate frequently when drinking alcoholic drinks. This dehydrates the body, producing headache and nausea—a hangover!

The kidneys are found between the peritoneum (the membrane of the abdominal cavity) and the back of the body. They receive oxygenated blood from the renal artery and drain into the renal vein. Each of the kidneys is a little bigger than a fist and is divided into two regions: the cortex and the medulla. In the cortex, approximately 1.2 million filtering units called nephrons filter the blood, forming urine that drains into the regions of the medulla (the renal pyramids, the minor calyx, the pelvic space, and the major calyx) before leaving the kidneys via the renal pelvis and the ureter, which drains into the bladder. As well as forming urine, the kidneys carry out a whole host of regulatory functions. They control long-term blood pressure via blood volume and regulate levels of acidity (pH)—if acidity levels change, this has an effect on all chemical reactions in the body. In addition, the kidneys excrete metabolic waste products and foreign substances, such as medicines, can make glucose during fasting, and produce and secrete hormones that regulate blood pressure, red blood cell production, and calcium balance for healthy bones.

RELATED TOPICS
See also
THE CIRCULATORY SYSTEM
page 62
THE MAJOR ARTERIES & VEINS
page 66
THE BLADDER
page 98

3-SECOND BIOGRAPHY
FRIEDRICH G. J. HENLE
1809–1885
A German physician, pathologist, and anatomist credited with the discovery of part of the renal tubule now known as the loop of Henle

30-SECOND TEXT
Marina Sawdon

We each have two kidneys, just under 4 inches (10 cm) long, in the upper-middle part of the back, but the body is able to function with only one kidney.

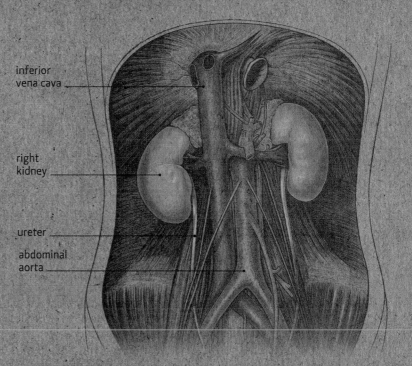

inferior
vena cava

right
kidney

ureter

abdominal
aorta

THE KIDNEYS IN SITU

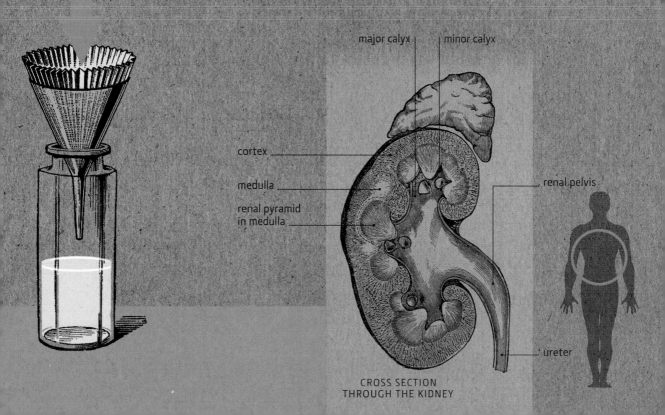

major calyx   minor calyx

cortex

medulla

renal pyramid
in medulla

renal pelvis

ureter

CROSS SECTION
THROUGH THE KIDNEY

THE BLADDER

the 30-second anatomy

RELATED TOPICS
See also
THE KIDNEYS
page 96
THE PELVIC FLOOR MUSCLES
page 146

30-SECOND TEXT
Gabrielle M. Finn

3-SECOND INCISION
The bladder is a saclike structure that receives and stores urine from the kidneys.

3-MINUTE DISSECTION
The urinary bladder holds between $1\frac{1}{4}$ and $2\frac{1}{2}$ cups (300–600 ml) of urine. As urine accumulates, folds (rugae) in the bladder wall flatten. The bladder wall thins as it stretches, letting it store increased volumes of urine. When a person urinates, the detrusor muscles contract to squeeze the urine out of the bladder and into the urethra. The process of urinating (micturition) is a reflex controlled by the central nervous system.

In infancy, the bladder sits within the abdomen. As a child reaches puberty, the bladder descends to its final position within the pelvis. The bladder is a hollow receptacle for urine with an inverted tetrahedral shape. Its muscle wall is formed by the detrusor muscle, which is highly distensible—it allows the bladder to swell. In adults, the bladder sits behind the pubic bone, and as it fills, it can extend as high as the navel (umbilicus). Urine drains from the kidneys into the bladder via two ureters, located in the upper corners of the bladder in an area known as the trigone, and leaves the bladder via the urethra, which is beneath the neck of the bladder and exits the body through the pelvic floor. The male urethra is about 8 inches (20 cm) long and travels to the tip of the penis; the female urethra is about $1\frac{1}{2}$ inches (4 cm) long and exits the body at the external urethral orifice, between the female labia.

The bladder can move fairly freely within the pelvis, apart from at its neck (at the base of the bladder), where it is held in place by ligaments and fascia.

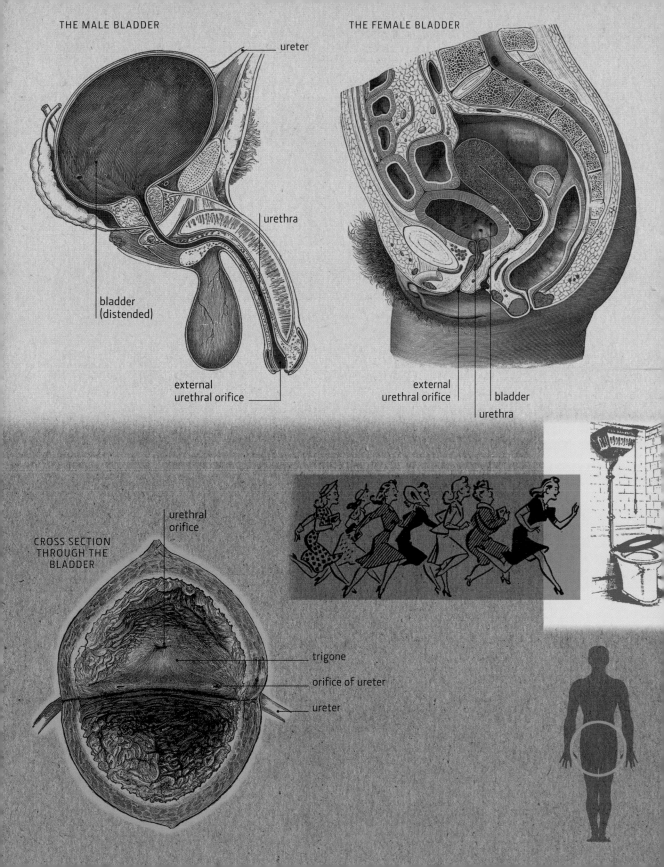

THE MALE BLADDER

ureter

urethra

bladder
(distended)

external
urethral orifice

THE FEMALE BLADDER

external
urethral orifice

bladder

urethra

CROSS SECTION
THROUGH THE
BLADDER

urethral
orifice

trigone

orifice of ureter

ureter

THE LYMPHATIC SYSTEM

the 30-second anatomy

3-SECOND INCISION
The lymphatic system is a series of tubes with nodules that collect fluid, proteins, and fats and deliver them to the blood circulatory system.

3-MINUTE DISSECTION
The lymph nodes tend to become swollen when the body is fighting an infection—for example, if a person has a bacterial infection, such as a sore throat, or a viral infection, such as glandular fever. Doctors examine swollen lymph nodes carefully because sometimes the swelling indicates cancer.

Some fluid and proteins leak out of blood vessels and would build up, causing swelling (edema) in the body, if there were not a way for them to be returned to the blood. The lymphatic system collects these fluids, proteins, and fats from the digestive system. The lymphatic system starts as very small, blind-ended tubes that are found virtually everywhere in the body close to the smallest blood vessels. These small tubes (lymphatic vessels) have gaps in them that allow for fluid, proteins, and even bacteria to enter; the fluid inside the lymphatic vessels is known as lymph. Lymph moves through the lymphatic system mainly by means of the vessels being squashed when muscles contract; the presence of one-way valves in the tubes means that the fluid is squeezed only in one direction. Bean-shaped nodules or nodes (lymph nodes) packed full of white blood cells are found at intervals along the lymphatic system. These white blood cells fight infection and serve to protect the body. The smallest vessels then combine to form larger lymph vessels and eventually form lymphatic ducts that drain lymph into the subclavian veins in the upper chest to mix with the blood.

RELATED TOPICS
See also
THE HEART
page 64
THE MAJOR ARTERIES & VEINS
page 66
THE MICROCIRCULATION
page 68

3-SECOND BIOGRAPHY
JOSEF ROTTER
1857–1924
A German surgeon who described small interpectoral lymph nodes that now bear his name. They receive lymphatic fluid from the muscles and mammary gland and are associated with breast cancer

30-SECOND TEXT
Andrew T. Chaytor

In addition to attacking bacteria and other agents of infection, lymph glands also perform the key function of filtering lymph.

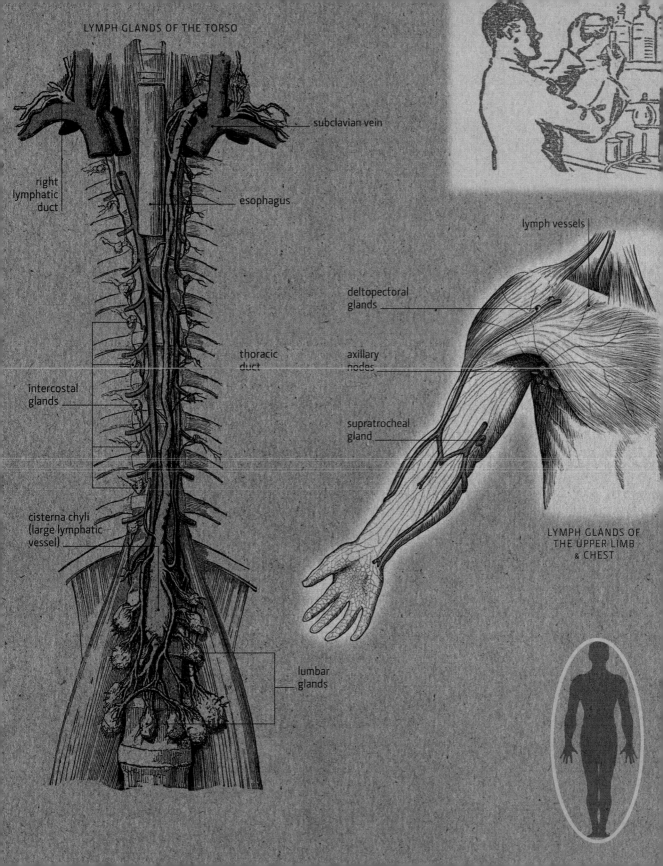

LYMPH GLANDS OF THE TORSO

subclavian vein

right
lymphatic
duct

esophagus

thoracic
duct

deltopectoral
glands

lymph vessels

axillary
nodes

intercostal
glands

supratrocheal
gland

cisterna chyli
(large lymphatic
vessel)

LYMPH GLANDS OF
THE UPPER LIMB
& CHEST

lumbar
glands

THE SENSORY & SPEECH ORGANS

auditory ossicles The three bones—malleus (hammer), incus (anvil), and stapes (stirrup)—that occupy the cavity of the middle ear and whose function is to transmit sound from the tympanic membrane (eardrum) to the inner ear, converting movements of air that strike the eardrum into physical movements that affect fluid in the inner ear.

auricle External part of the ear, consisting of cartilage covered by skin, also known as pinna; the earlobe is made of fat. The inner opening (concha) gives into the external auditory canal, which runs for 1 inch (2.5 cm) to the eardrum. This whole structure is called the outer ear. The function of the auricle is to direct sound into the auditory canal.

cochlea Hollow tube shaped like the shell of a snail in the inner ear. The cochlea contains the organ of Corti, in which rods and filaments convert sound vibrations from the middle ear into electrical impulses in the acoustic nerve; these impulses are transmitted to the brain, where they are interpreted as sounds. The cochlea is the part of the inner ear with a function in hearing; the other parts are concerned with balance.

cornea Transparent section on the central front part of the outer eye through which light enters. The cornea connects to the sclera at the limbus.

dermatome An area of skin supplied by a single spinal nerve; each dermatome is named after the nerve that supplies it. For instance, the dermatome T5 is supplied by the fifth thoracic nerve (the fifth nerve emerging from the thoracic vertebrae). Anatomy textbooks represent dermatomes as colored bands.

epidermis Outer layer of the body's skin, beneath which—in the basal cell layer—new skin cells are made. The epidermis is just $1/250$ inch (0.1 mm) thick on the eyelids but up to $1/25$ inch (1 mm) thick on the soles of the feet and palms of the hands; it contains Langerhans cells that are part of the body's immune system.

iris The colored section of the eye, a muscular diaphragm that dilates or constricts its central opening (the pupil) in order to let more or less light into eye.

lens Elastic transparent round section about $2/5$ inch (10 mm) in diameter behind the iris in the eye; the lens brings light to a focus on the retina. In between the lens and the retina lies a cavity containing jellylike vitreous body.

nasal conchae Three projections of turbinate bone in the lateral wall of the nasal cavity: the superior, middle, and inferior conchae. The conchae create four passages within the nose and force inhaled air into a steady flow.

olfactory area Region at the top of the nasal cavity that contains up to 25 million olfactory receptors sensitive to airborne compounds; nerve signals sent by the receptors are processed in the brain as smells.

papillae Protuberances on the upper surface of the tongue, of which there are three types: fungiform, circumvallate, and filiform papillae. Fungiform and circumvallate papillae contain taste buds, which enable the tongue to perform its function as an organ of the sense of taste. Filiform papillae do not contain taste buds and their function is mechanical.

pharynx Muscular tube leading from the nasal cavity and mouth to the esophagus and larynx (voice box). Two tubes (the Eustachian tubes) link the pharynx to the middle ear.

retina Membrane in the inner layer of the eyeball, containing light-sensitive photo-receptor cells called cones and rods. In the retina, there are about 100 million rods (specialized for low light, movement, and black-and-white vision) and 6 million cones (functioning best in bright and medium light, fine detail, and color vision). Signals from the retina pass down the retinal nerve to the optic disk, and from there along the optic nerve to two visual cortices, which are situated one on each side of the brain.

sclera The opaque main part of the outer eye (commonly known as the "white of the eye").

THE DERMATOMES

the 30-second anatomy

In medical textbooks, the dermatomes are depicted as colored bands covering the human body—although in reality they are invisible. A dermatome is a region of skin supplied by sensory fibers from a single spinal nerve. The name of each dermatome is derived from that of the spinal nerve by which it is supplied—for example, the area of skin supplied by the seventh cervical spinal nerve is known as dermatome C7. The spinal nerves emerge on each side of the body from the midline (an imaginary line that splits the body into two). From these nerves, the dermatomes occur separately in a symmetrical pattern on the right and left sides of the body. The dermatome pattern follows the spinal nerves: cervical nerves supply the head and upper limbs, thoracic nerves supply the trunk and parts of the upper limb, and lumbosacral nerves supply the pelvis and lower limbs. Some diseases cause a rash across one or more specific dermatomes—the area in which the disease spreads reveals its neurological origin.

3-SECOND INCISION
A dermatome is a band or area of skin supplied by sensory fibers from a single spinal nerve.

3-MINUTE DISSECTION
Dermatomes are used clinically to identify the origin of neurological diseases. If a patient feels numb from the waist down, a physician could test the sensation in the skin moving steadily down the thorax and abdomen. Once the physician finds the point at which sensation is lost, he or she can determine which dermatome is affected and which spinal nerve is injured.

RELATED TOPICS
See also
THE SPINAL CORD
page 130

3-SECOND BIOGRAPHY
SIR HENRY HEAD
1861–1940
An English neurologist who conducted the original research to map dermatomes

30-SECOND TEXT
Gabrielle M. Finn

Key dermatomes include T10, which supplies (innervates) nerve fibers to the skin around the navel and T4, which innervates the area of the nipples.

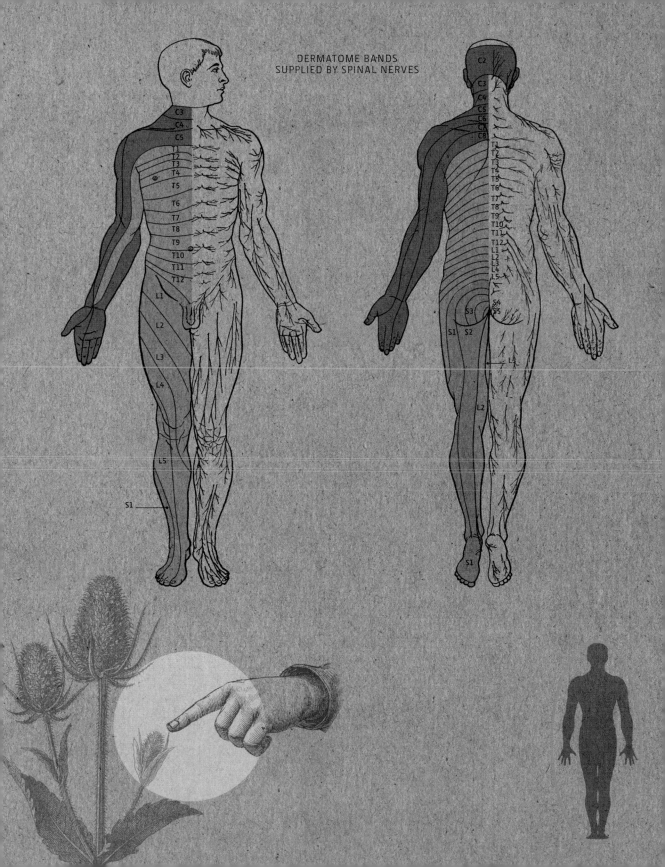

DERMATOME BANDS
SUPPLIED BY SPINAL NERVES

THE SKIN, HAIR & NAILS

the 30-second anatomy

Skin consists of three main

layers—the outer epidermis; the dermis, where the blood vessels, nerves, glands, and supportive tissues are situated; and the hypodermis, where the protective and insulating layer of fat can be found. Cells in the epidermis are constantly replacing themselves, taking around 26 days to do so. Hair, which grows from follicles in the dermis, is found on almost every surface of the body, and its role is to trap a layer of warm air next to the skin to maintain body temperature. On the end of fingers and toes, nails are hard, fused plates of keratin, thin enough to let light pass through and reflect off the pink nail bed underneath; the pale semicircle visible at the base of the nail is known as the lunula, and is the area from which the nail develops and grows. Nails grow at a rate of about $\frac{1}{10}$ inch (3 mm) per month, although fingernails grow faster than toenails. Thickened and discolored nails are a sign of slow growth, indicating a health problem.

3-SECOND INCISION
Skin is a protective insulator that synthesizes vitamin D, hair helps maintain body temperature, while nails provide a degree of protection and on fingers help in picking things up.

3-MINUTE DISSECTION
Color in skin and hair is derived from the protein melanin. Melanin can be either blue-black (eumelanin) or red-brown (phaeomelanin); relative levels of these account for the differing shades of skin and hair. Without melanin, hair is white—as occurs naturally with aging. It is unusual to have skin without melanin, but this does happen in albinism; individuals with this condition also have no pigment in their eyes. Melanin is not present in fingernails or toenails.

RELATED TOPICS
See also
THE DERMATOMES
page 106

3-SECOND BIOGRAPHY
GASPARE TAGLIACOZZI
1545–1599
An Italian surgeon who developed the Indian process of skin grafting, correctly anticipating that the use of foreign body tissue would be rejected in the transplantation

30-SECOND TEXT
Judith Barbaro-Brown

The protein keratin that makes up skin, hair, and nails comes in different forms—so hair is flexible, skin elastic, and nails hard.

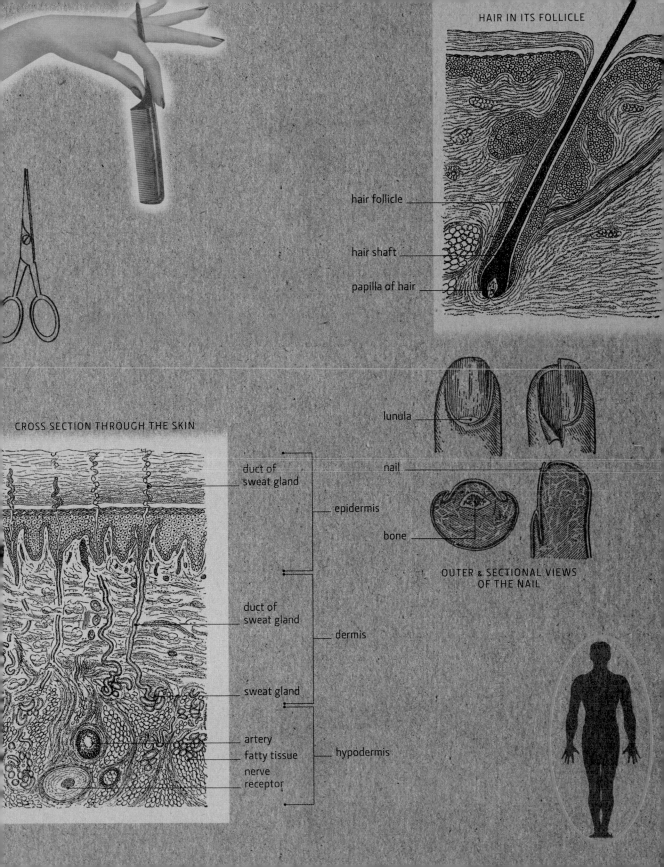

HAIR IN ITS FOLLICLE

hair follicle

hair shaft

papilla of hair

CROSS SECTION THROUGH THE SKIN

duct of
sweat gland

epidermis

duct of
sweat gland

dermis

sweat gland

artery
fatty tissue
nerve
receptor

hypodermis

lunula

nail

bone

OUTER & SECTIONAL VIEWS
OF THE NAIL

THE EYES

the 30-second anatomy

Each eye is well protected—it sits in a bony socket (orbit) formed by the bones of the skull, while eyelashes, eyebrows, and eyelids prevent foreign materials from entering. Within the eyelid, a thin tissue called conjunctiva produces mucus that combines with tears from the lacrimal glands to lubricate the eye. The eyeball has three layers: the outer sclera, or "white of the eye," is tough with no blood vessels; the choroid coat is delicate, darker, contains several blood vessels, and prevents light from scattering; the inner retina contains light-sensitive rods and cones. The iris is the colored part of the eye; its central opening is called the pupil. The amount of light that enters the eye is controlled by circular muscles (ciliary muscles) that make the pupil narrower in bright light and others resembling spokes of a wheel that pull the opening wider in dim light. After incoming light passes through the clear cornea, it is focused by the lens onto a small area of the retina. Signals from the photoreceptor cells of the retina are carried to the brain by the optic nerve. Aqueous humor maintains the shape of the cornea while the jellylike vitreous body supports the eye behind the lens. Six extrinsic muscles produce coordinated eye movements.

3-SECOND INCISION
The eyes are specialized sight organs—they focus light on the sensitive retina and then dispatch electrochemical impulses along the optic nerve to the brain's visual cortices.

3-MINUTE DISSECTION
The reason a person reaches for a tissue when crying is that tears from the eyes drain into the nose. Tears contain an enzyme that helps prevent infection, but even a tiny scratch on the eye could possibly lead to blindness. It is very important to protect the eyes.

RELATED TOPICS
THE SKULL
page 22
THE CRANIAL NERVES
page 136

3-SECOND BIOGRAPHY
SIR HAROLD RIDLEY
1906–2001
An English ophthalmic surgeon who in 1949 performed the first cataract operation

30-SECOND TEXT
Jo Bishop

In some people, the lens becomes clouded, causing severely reduced vision; it is called a cataract and mostly arises as an effect of aging.

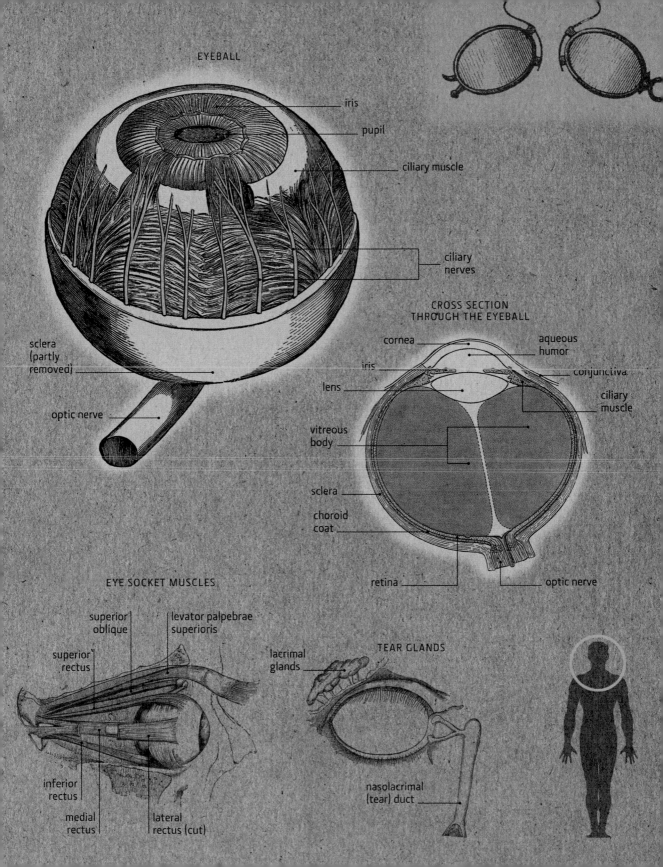

EYEBALL

iris

pupil

ciliary muscle

ciliary nerves

sclera (partly removed)

optic nerve

CROSS SECTION THROUGH THE EYEBALL

cornea

aqueous humor

iris

conjunctiva

lens

ciliary muscle

vitreous body

sclera

choroid coat

retina

optic nerve

EYE SOCKET MUSCLES

superior oblique

levator palpebrae superioris

superior rectus

lacrimal glands

TEAR GLANDS

inferior rectus

medial rectus

lateral rectus (cut)

nasolacrimal (tear) duct

THE NOSE

the 30-second anatomy

The external nose varies in size and shape. Air normally enters the body's respiratory system through the nostrils, passing into two nasal cavities separated by a nasal septum formed of bone and cartilage. Nasal hairs protect the entrance (vestibule) of the nose, trapping foreign matter, such as dust particles. Incoming air is warmed up and moistened by flowing through narrow passageways (meati) that contain several blood vessels and run between three shell-like projections of the nasal cavity's lateral wall (the superior, middle, and inferior conchae). The conchae are covered by a sticky mucous membrane; mucus also drains from four paranasal air sinuses (frontal, maxillary, ethmoidal, and sphenoidal) into the nasal cavity—and is passed back to the throat, where it is swallowed. These safeguards protect the delicate exchange surface of lung tissue in the lower portions of the respiratory tract from cold air and foreign particles. Paired olfactory organs in the upper part of the nasal cavity contain receptors stimulated when airborne compounds dissolve. The olfactory nerve carries this information to the brain's olfactory bulb, which processes the nerve impulses as smells.

RELATED TOPICS
See also
THE LUNGS
page 76
THE BRONCHIAL TREE
page 78

3-SECOND BIOGRAPHY
SIR WILLIAM BOWMAN
1816–1892
An English surgeon and anatomist, who named the Bowman's glands within the olfactory mucosa

30-SECOND TEXT
Jo Bishop

Because it protrudes from the face and its bones are fragile, the nose is vulnerable to injury—almost four out of ten facial injuries are broken noses.

CROSS SECTION THROUGH THE NASAL CAVITY

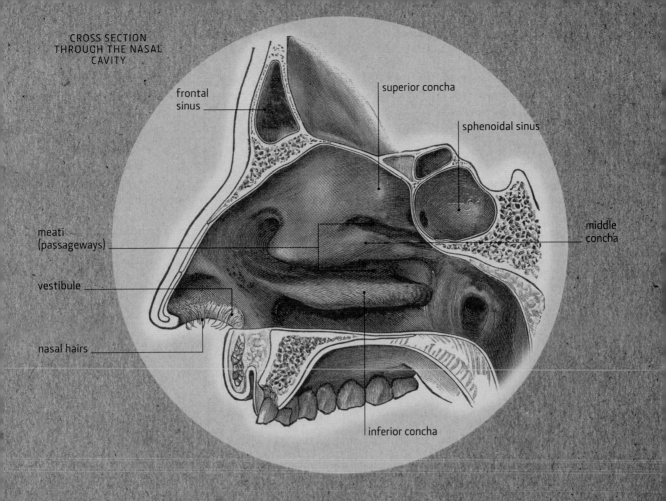

superior concha

frontal sinus

sphenoidal sinus

middle concha

meati (passageways)

vestibule

nasal hairs

inferior concha

NASAL CARTILAGES

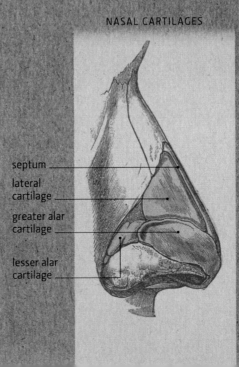

septum

lateral cartilage

greater alar cartilage

lesser alar cartilage

ca. 129 CE
Born in Pergamon
(modern Bergama in
Turkey)

143
Began studying
philosophy

145
Sent to study medicine
after his father dreamed
of Asclepius, the god of
healing

148
Following the death of
his father, traveled and
studied in Alexandria
and Smyrna

157
Became physician at the
gladiator school in
Pergamon

161
Moved to Rome

166
Forced to flee the city
after threats from
enemies

168
Invited back to Rome as
the physician to Marcus
Aurelius; became, in turn,
physician to Commodus
and Septimius Severus

216/217
Died in Rome

GALEN

Aelius Galenus or Claudius Galenus, known as Galen, is the anatomists' anatomist. A Greek philosopher-physician from a wealthy family, he trained at the prestigious medical schools of Alexandria and Smyrna (among others), and spent most of his professional life in Rome at the imperial court; Roman Emperor Marcus Aurelius said of him *primum sane medicorum esse, philosophorum autem solum* (he was "first among doctors and unique among philosophers"). Considered revolutionary in second-century Rome, Galen's theories, practice, and approach became hard and fast orthodoxy—so much so that when they were challenged more than a millennium later by Renaissance anatomists, this was regarded as near blasphemy by the medical establishment.

Galen advanced the theories of the Greek physician Hippocrates (ca. 450–ca. 370 BCE)—ideas that were not always well received, because they included bloodletting—and pursued a radical, hands-on approach to anatomy that caused controversy. An early stint as physician in residence at the School of Gladiators in his home town of Pergamon allowed for him to study the form and function of the human body while he tended the wounds and oversaw the training and nutrition of his charges; however, because Rome had banned human dissection in 150 BCE, when he wanted to examine internal organs, he had to use the bodies (sometimes still living) of pigs and primates—chosen because they are the nearest in form to human anatomy; this explains why some of his conclusions were erroneous. He was the originator of many modern medical practices (such as those of making a diagnosis and a giving a prognosis) but is remembered above all as an anatomist for his work on the trachea and the circulatory system.

Ambitious, rich, clever, and successful, Galen was also prolific—and produced an estimated 600 treatises. His early training in humanities, particularly in comparative philosophy, enabled him to break free from following any particular school and gave him a unique take on the body of medical knowledge. Unfortunately, much of his work has been lost, destroyed, or confused, and there were forgeries and "bootleg" editions of his work made in his own lifetime as well as later; a good deal of the surviving work takes the form of translations of translations, so it has been difficult to establish a reliable canon or to date his work accurately.

THE EARS

the 30-second anatomy

The part of the ear visible on the side of the skull is called the auricle. This is a flexible, skin-covered curl of cartilage that helps direct sound waves into the 1 inch (2.5 cm) external auditory canal; this passage leads to the eardrum (tympanic membrane). These elements are all part of the external ear. The middle ear consists of a tympanic cavity connected to the throat (nasopharynx) by a 1½ inch (4 cm) passageway called the auditory tube that functions to equalize the pressure in the middle ear with outside pressure. Three small bones, the auditory ossicles—malleus (hammer), incus (anvil), and stapes (stirrup)—transfer sound vibrations from the tympanic membrane to a fluid-filled chamber, the membranous labyrinth. The labyrinth protects the important inner ear structures; it is divided into three parts and filled with perilymph. The first two parts—the vestibule and three projecting bony canals—contain balance receptors. The third part, the cochlea, is coiled like a snail and contains hearing receptors in the organ of Corti. The vestibular nerve (nerve fibers carrying information to the brain from the vestibule and canals) joins the cochlear nerve (carrying information from the cochlea) to form the vestibulocochlear nerve.

RELATED TOPICS
THE CRANIAL NERVES
page 136

3-SECOND BIOGRAPHY
MARQUIS ALFONSO GIACOMO GASPARE CORTI
1822–1876
An Italian anatomist who carried out microscopic research into the cochlea

30-SECOND TEXT
Jo Bishop

3-SECOND INCISION
The specialized sense organ for hearing and balance, the ear is divided into three parts—external, middle, and inner ear.

3-MINUTE DISSECTION
Children are prone to ear infections because their auditory tubes are shorter and broader than adults, letting infections travel from the throat. The auditory tube can be forced open by swallowing hard, blowing your nose with your mouth closed, or yawning—this is normally done to relieve pressure pain caused during flying.

Wax and hairs in the external auditory canal of the outer ear prevent foreign bodies from entering and work against infection.

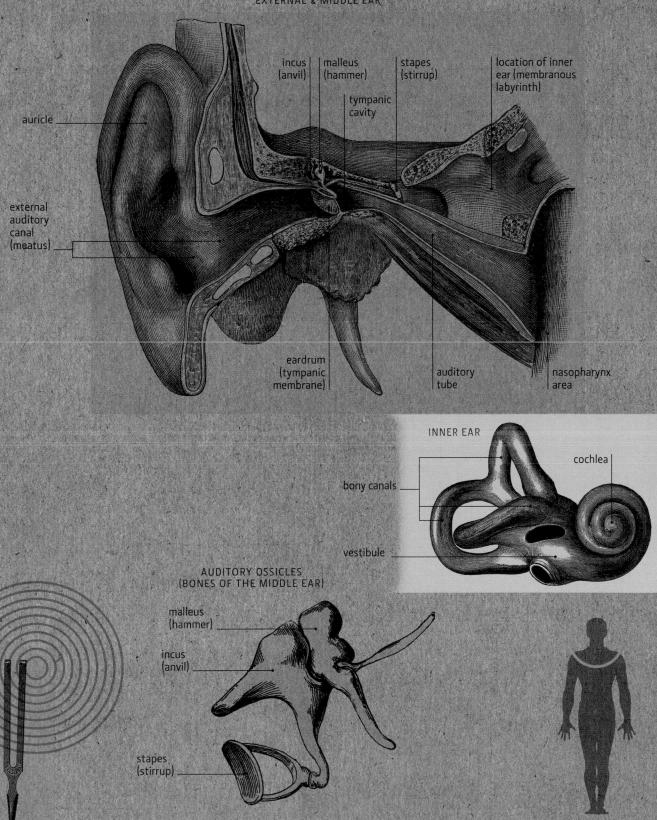

EXTERNAL & MIDDLE EAR

incus
(anvil)

malleus
(hammer)

stapes
(stirrup)

location of inner
ear (membranous
labyrinth)

tympanic
cavity

auricle

external
auditory
canal
(meatus)

eardrum
(tympanic
membrane)

auditory
tube

nasopharynx
area

INNER EAR

cochlea

bony canals

vestibule

AUDITORY OSSICLES
(BONES OF THE MIDDLE EAR)

malleus
(hammer)

incus
(anvil)

stapes
(stirrup)

THE TONGUE

the 30-second anatomy

The front part of the tongue is its body, situated within the oral cavity, and the back part is its root, at the back of the throat. The tongue's upper surface or dorsum contains small projections (papillae). There are three types of papillae, found in distinct areas of the tongue: fungiform papillae are round, with taste buds along their sides; circumvallate papillae, the largest, form a V shape near the border between the front and rear of the tongue; cone-shaped filiform papillae, are the smallest and do not contain taste buds. Information from the taste receptors is carried to the brain by the facial nerve and the glossopharyngeal nerve. The underside of the tongue is thin and delicate, and it is connected to the floor of the mouth by the frenulum, which anchors the tongue and restricts its movements. The lingual tonsil is found on the tongue's root at the back. Most of the tongue is made up of muscle; it has extrinsic (external) muscles responsible for large tongue movements and intrinsic (internal) muscles that alter the tongue's shape and operate with the extrinsic tongue muscles during delicate and precise movements, such as speaking. Most of the muscles are activated by the hypoglossal nerve.

RELATED TOPICS
See also
THE NOSE
page 112
THE CRANIAL NERVES
page 136

3-SECOND INCISION
The tongue is a piece of muscular tissue used in chewing, swallowing, and speaking—it contains taste buds with receptors for sweet, salty, sour, and bitter tastes.

3-MINUTE DISSECTION
Each adult has around 10,000 taste buds. Alongside the primary taste sensations of sweet, salty, sour, and bitter, a fifth has been described—umami (a savory taste). The sense of smell is involved with how people taste things—this is why food tastes dull if a person has a cold. Biting the tip of the tongue promotes the production of extra saliva.

3-SECOND BIOGRAPHY
DAVID HÄNIG
practicing in 1900s
A German physician who in 1901 made the first description of the tongue's "taste map"

30-SECOND TEXT
Jo Bishop

There are more filiform than other types of papillae on the tongue; they do not contain taste buds and so play no part in tasting.

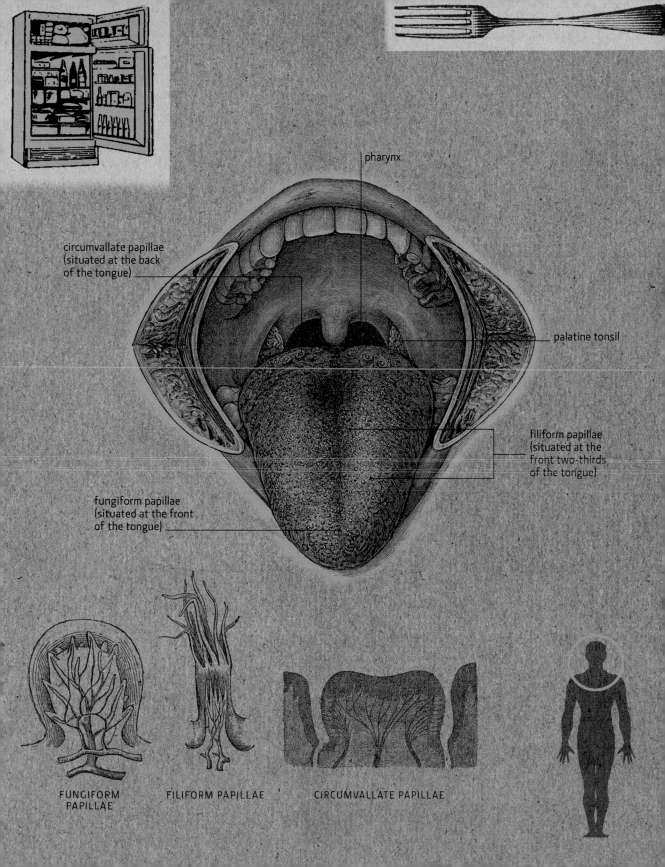

pharynx

circumvallate papillae
(situated at the back
of the tongue)

palatine tonsil

filiform papillae
(situated at the
front two-thirds
of the tongue)

fungiform papillae
(situated at the front
of the tongue)

FUNGIFORM
PAPILLAE

FILIFORM PAPILLAE

CIRCUMVALLATE PAPILLAE

THE PHARYNX, LARYNX & VOCAL CORDS

the 30-second anatomy

3-SECOND INCISION
The pharynx links the mouth to the digestive and respiratory systems, while the larynx—situated between the pharynx and trachea (windpipe)—houses the vocal cords.

3-MINUTE DISSECTION
Children's voices have a higher pitch than those of adults because children's vocal folds are thinner and shorter. Men have deeper voices because during puberty the larynx enlarges and the vocal cords become thicker and longer. When a person is singing a high note, the vocal folds can oscillate (vibrate) at 440 times per second.

The pharynx, or throat, begins behind the nasal cavity as the nasopharynx, and continues behind the mouth as the oropharynx. Its lowest portion is the laryngopharynx, which becomes the larynx at the front and the esophagus behind. The larynx, or voice box, is made of three unpaired cartilages (thyroid and cricoid cartilages, and the epiglottis) and three pairs of smaller cartilages (arytenoid, corniculate, and cuneiform cartilages). Of these the largest is the thyroid cartilage: it protrudes from the neck and is known as the Adam's apple, and it is larger and more prominent in men than women. The cartilages are held together by intrinsic (internal) ligaments and attached to the surrounding structures by extrinsic (external) ligaments. The vestibular ligaments lie within vestibular folds (false vocal cords) between the thyroid and arytenoid cartilages and serve, along with the epiglottis, to prevent foreign objects from entering the trachea. The vocal folds containing vocal ligaments (true vocal cords) are involved in the production of sound. They vibrate when air from the lungs passes through the gap between them (glottis). The type of sound produced is dependent on the diameter, length, and tension of the vocal folds.

RELATED TOPICS
See also
THE LUNGS
page 76
THE BRONCHIAL TREE
page 78

3-SECOND BIOGRAPHY
BENJAMIN GUY BABINGTON
1794–1866
An English physician who invented the first laryngoscope (a tube used to look down the larynx)

30-SECOND TEXT
Jo Bishop

The pharynx runs from the base of the skull to the level of the sixth cervical vertebra and is 5 inches (12.5 cm) in length.

CROSS SECTION THROUGH THE SPEECH ORGANS

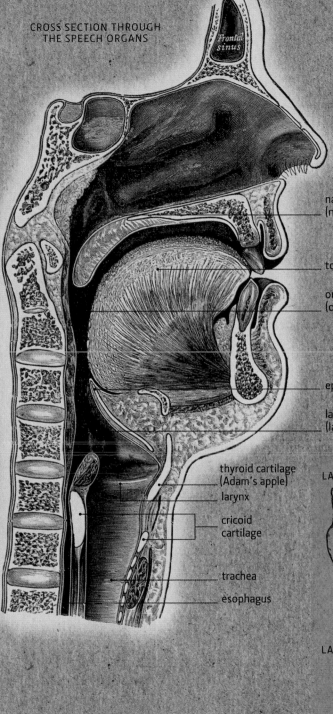

Frontal sinus

nasopharynx (nasal part of pharynx)

tongue

oropharynx (oral part of pharynx)

epiglottis

laryngopharynx (laryngeal part of pharynx)

thyroid cartilage (Adam's apple)

larynx

cricoid cartilage

trachea

esophagus

LARYNX: INTERIOR VIEW

cricoid cartilage

arytenoid cartilage

vocal ligaments (vocal cords)

thyroid cartilage

LARYNX: EXTERIOR VIEW

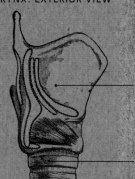

thyroid cartilage

trachea

THE ENDOCRINE & NERVOUS SYSTEMS

THE ENDOCRINE & NERVOUS SYSTEMS
GLOSSARY

adrenal glands Pair of glands, one on each kidney, that are part of the endocrine system of organ and glands that secrete hormones into the blood to regulate bodily activities. The adrenal glands release hormones controlling how the body uses food, as well as moderating blood pressure and heart rate.

axon Long tendril emanating from the body of a neuron, along which nerve impulses travel at a speed of around 325 feet (100 m) per second. Some axons are encased in a sheath of a fatty substance called myelin, which insulates the axon and speeds transmission.

basal ganglia Four areas of gray matter in the lower part of the brain's cerebral hemispheres.

brain stem Region in the lower rear of the brain that connects the middle and upper-forward parts of the brain to the spinal cord. With the cerebellum, the brain stem is part of the hindbrain.

cerebellum Section in the lower rear of the brain connecting the brain stem to the rear of the cerebrum. It has a role in maintaining balance and muscle coordination.

cerebral cortex Wrinkled upper surface of the cerebrum. The cerebral cortex controls most thought processes and contains the bodies of neurons (nerve cells). These connect via long tendrils or axons.

cerebrum Brain's largest part, divided into two halves—the left and right hemispheres—that are identical in appearance. Each hemisphere contains four regions or lobes: the frontal (front), occipital (rear), temporal (side), and parietal (top) lobes. The two halves are connected by a band of nerve fibers—the corpus callosum.

cervical nerves Spinal nerves issuing from the cervical vertebrae (the seven bones of the neck, at the upper end of the backbone); sensory fibers from these nerves run to the head and upper limbs.

cranial nerves Nerves that have their origin in the brain stem and related brain structures, and which supply the brain, head, and neck.

dura mater One of the three meninges (connective tissue coverings) that enclose and protect the spinal cord and brain. The dura mater is the outer covering, and the thickest of the three meninges. It contains the arachnoid mater and pia mater; between these layers cerebrospinal fluid fills the subarachnoid space.

glial cell Type of cell in the connective tissue of the brain and nervous system. Glial cells support neurons by providing nourishment and removing waste matter.

gray matter Tissue in the brain and spinal cord, brownish gray in color, consisting mainly of the cell bodies of neurons.

hypothalamus Brain region below the thalamus and above the brain stem. It releases neurohormones that regulate the secretion of hormones by the pituitary gland and other glands of the endocrine system.

lumbosacral nerves Spinal nerves issuing from the lumbar vertebrae (five vertebrae in the lower back) and sacrum (the lowest part of the spinal column); these nerves supply the pelvis and lower limbs.

neuron Nerve cell in the brain and nerves. The bodies of neurons communicate along tendrils called axons. When axons meet, they do not touch; impulses are carried across the gap between the axons (called the synaptic cleft) by neurotransmitter chemicals.

thoracic nerves Spinal nerves issuing from the thoracic vertebrae (the 12 bones that form the middle of the vertebral column, between the cervical vertebrae of the neck and the lumbar vertebrae of the lower back); these nerves run to the trunk of the body and parts of the upper limbs.

thyroid Gland of the endocrine system. Situated in the neck, the thyroid releases the thyroid hormone that controls metabolism.

white matter Tissue in the brain and spinal cord (white in color) consisting mainly of glial cells and nerve fibers (axons) connecting the cell bodies of nerve cells (neurons).

THE ENDOCRINE SYSTEM

the 30-second anatomy

3-SECOND INCISION

The endocrine system controls body processes through secretions of hormones by the hypothalamus, pituitary, parathyroid, thyroid, pancreatic, and adrenal glands, testis, and ovaries.

3-MINUTE DISSECTION

The thyroid hormones T3 and T4 help regulate the speed of activity of all cells of the body. For this reason, disorders of the thyroid gland may manifest as nonspecific and generalized problems—such as, perhaps, puffy skin, hoarseness, lack of appetite, disturbances in the menstrual cycle, sensitivity to the cold, or vision/hearing problems. In hypothyroidism, the thyroid gland is insufficiently active; in the less common hyperthyroidism it is overactive.

The endocrine system regulates body functions, such as development and growth, speed of cell function, muscle tone, and water and electrolyte balance. Its organs are very diverse. They are distinct endocrine glands (pituitary, thyroid, parathyroid glands), components of other organs (kidneys, ovary, testis), or diffuse endocrine glands (in the digestive tract). Endocrine glands are usually solid organs made up of very active cells organized around an abundance of blood vessels and some loose supporting tissue. These cells synthesize and secrete chemical materials called hormones into the blood, which transports them to the places at which they function. Some hormones act in chains, causing the release of another hormone, and are called either releasing (first-order) or stimulating (second-order) hormones. Thyroxine, which is secreted by the thyroid gland, located below the Adam's apple in the front of the neck, is an example. The thyroid gland secretes thyroxine when the thyroid-stimulating hormone produced by the pituitary gland acts on it. The pituitary gland is itself first stimulated by a thyrotropin-releasing hormone that is produced by the hypothalamus.

RELATED TOPICS

See also
THE PANCREAS
page 92
THE FEMALE REPRODUCTIVE SYSTEM
page 144
THE MALE REPRODUCTIVE SYSTEM
page 150

3-SECOND BIOGRAPHY
EMIL THEODOR KOCHER
1841–1917
A pioneering Swiss surgeon who performed more than 5,000 thyroid gland operations

30-SECOND TEXT
December S. K. Ikah

Endocrine organs are widely dispersed through the body— from brain to digestive tract to reproductive system.

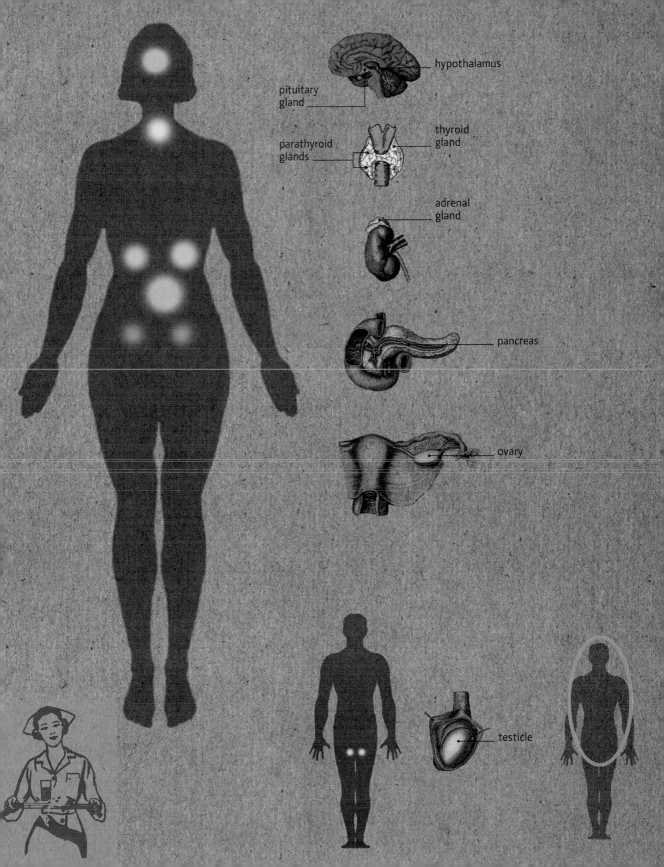

hypothalamus

pituitary
gland

parathyroid
glands

thyroid
gland

adrenal
gland

pancreas

ovary

testicle

THE BRAIN
& BRAIN STEM

the 30-second anatomy

The brain is divided into three

regions—the forebrain, midbrain, and hindbrain (continuous with the spinal cord). The forebrain consists of the cerebrum, the basal ganglia, and the thalamus; the midbrain connects the forebrain and hindbrain; and the hindbrain is the brain stem and cerebellum. In the forebrain is the cerebrum, the brain's largest part, which is divided into two hemispheres, right and left. Its wrinkled upper surface is the cerebral cortex, which consists mainly of the cell bodies of nerve cells (neurons)—known as "gray matter" from their color; the neurons connect via long fibers called axons and branching dendrites—together known as "white matter." The other parts of the forebrain are: the basal ganglia (four areas of gray matter in each hemisphere) and the thalamus, a gateway for sensory fibers going to the cerebral cortex. In the hindbrain, the cerebellum connects the brain stem to the cerebrum and has a role in maintaining balance and coordination; the brain stem lies between the spinal cord, cerebellum, and forebrain—it connects those regions of the brain, transmits motor and sensory fiber tracts, contains cranial nerves, and controls breathing and circulation.

3-SECOND INCISION
The brain is the body's control center—it consists of brain stem, cerebellum, thalamus, basal ganglia, and cerebral hemispheres, enclosed in meninges within the skull.

3-MINUTE DISSECTION
A man's brain weighs on average 3 pounds (1.5 kg) and a woman's 2¾ pounds (1.25 kg)—a difference due to the typical variation in body size between men and women. This is about one-fiftieth of an adult's body weight. Yet the brain uses one-fifth of the lungs' oxygen supply.

RELATED TOPICS
THE SPINAL CORD
page 130
THE AUTONOMIC NERVOUS SYSTEM
page 134
THE CRANIAL NERVES
page 136

3-SECOND BIOGRAPHY
HIPPOCRATES ASCLEPIADES
460–377 BCE
An ancient Greek physician, celebrated as "the father of medicine," who first stated that the brain is involved in sensation and is the seat of human intelligence

30-SECOND TEXT
December S. K. Ikah

The two hemispheres of the cerebrum are divided by a cleft and connected at its bottom by a sheet of nerve fibers, the corpus callosum.

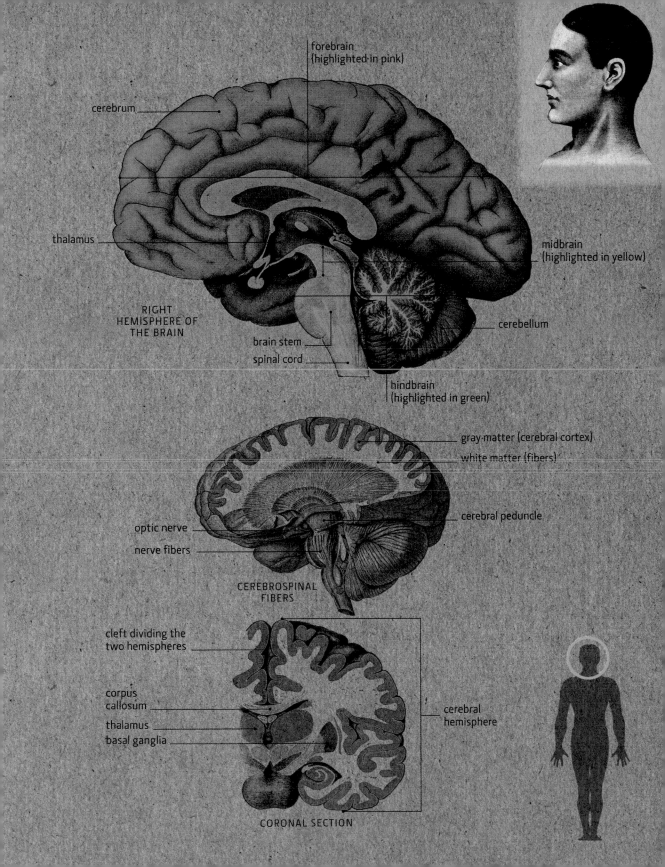

forebrain
(highlighted in pink)

cerebrum

thalamus

RIGHT
HEMISPHERE OF
THE BRAIN

midbrain
(highlighted in yellow)

cerebellum

brain stem

spinal cord

hindbrain
(highlighted in green)

gray matter (cerebral cortex)

white matter (fibers)

optic nerve

nerve fibers

cerebral peduncle

CEREBROSPINAL
FIBERS

cleft dividing the
two hemispheres

corpus
callosum

thalamus

basal ganglia

cerebral
hemisphere

CORONAL SECTION

THE SPINAL CORD

the 30-second anatomy

3-SECOND INCISION
The spinal cord is housed within the vertebral canal. It transmits signals to and from the brain in order to control movement, sensation, and reflexes.

3-MINUTE DISSECTION
The space between the arachnoid and pia maters (the subarachnoid space) is filled with cerebrospinal fluid, which can be sampled using a needle inserted around the lumbar region of the spine in a lumbar puncture. At this point, the spinal cord has split into fibrous strands (cauda equina), so inserting a needle does not damage the spinal cord. The fluid can be analyzed to look for illnesses or drained in order to relieve pressure around the brain.

The brain and spinal cord together form the central nervous system. The spinal cord lies within the upper two-thirds of the vertebral canal. It extends from the brain down as far as the second or third lumbar vertebrae. The spinal cord does not have a uniform diameter—along its length it exhibits two enlargements. The first is the cervical enlargement, which gives rise to the nerves that supply the upper limbs (brachial plexus); the second is the lumbosacral enlargement, which corresponds to nerves that supply the lower limbs. The functions of the spinal cord are to coordinate reflexes, to transmit motor (movement) signals down the cord from the brain, and to return sensory signals to the brain. Inside the cord is a central canal surrounded by gray matter and white matter, rich in nerve cell bodies and nerve cell processes that receive incoming signals. Three connective tissue coverings known as meninges surround the spinal cord—their role is to suspend and protect the cord within the vertebral canal. The outermost layer (dura mater) is the thickest; beneath this is the arachnoid mater and finally the pia mater, which is continuous with the both the brain and spinal cord.

RELATED TOPICS
See also
THE SPINE & RIB CAGE
page 24
THE BRAIN & BRAIN STEM
page 128
THE NERVE PLEXUSES
page 138

30-SECOND TEXT
Gabrielle M. Finn

At the base of the spinal column, the cauda equina is a bundle of nerve roots that connect to the legs, bladder, bowels, and genitals.

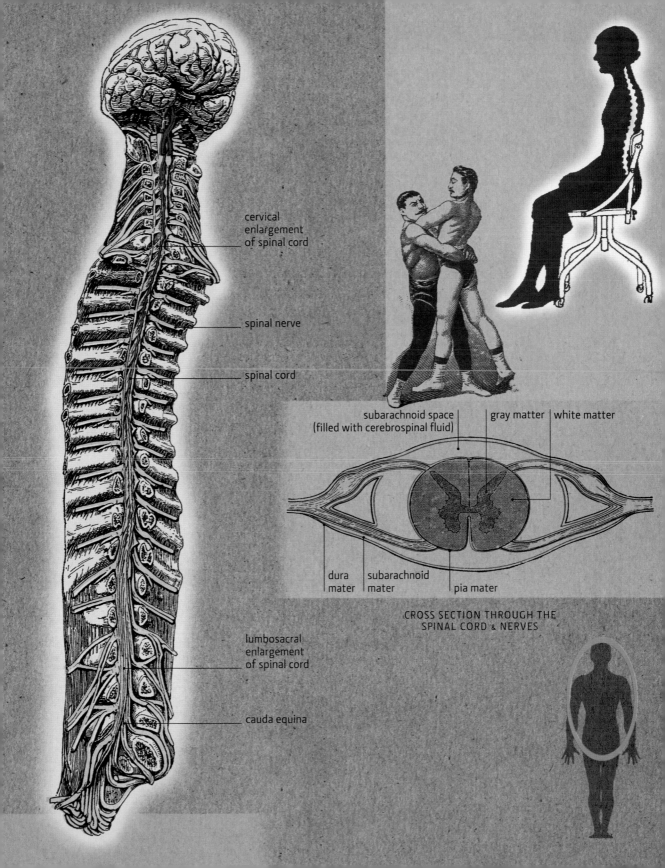

cervical
enlargement
of spinal cord

spinal nerve

spinal cord

lumbosacral
enlargement
of spinal cord

cauda equina

subarachnoid space
(filled with cerebrospinal fluid)

gray matter | white matter

dura
mater | subarachnoid
mater | pia mater

CROSS SECTION THROUGH THE
SPINAL CORD & NERVES

1827
Born in London

1845
Entered St. George's Hospital

1848
Became Member of the Royal
College of Surgeons

1849
Won the Triennial Prize of the
Royal College of Surgeons for an
essay on the optic nerves

1852
Elected Fellow of the Royal Society

1852
Presented a paper on "The Glands
of Chicks"

1853
Won the Astley Cooper Prize
(300 guineas) for his dissertation
on the spleen

1858
Published the first edition of
Anatomy Descriptive and Surgical
(later *Anatomy of the Human
Body*, and now *Gray's Anatomy*)

1860
Published the second edition of his
Anatomy Descriptive and Surgical

1861
Applied for the post of Assistant
Surgeon at St. George's Hospital

1861
Died of smallpox

HENRY GRAY

Gray's Anatomy **is probably the** best-known medical textbook in the Western world, an invaluable resource for the medical and artistic professions that is so indissolubly linked with its creator that its official title is now eponymous. However, Gray would have been far too modest to use his own name in the book's title in his lifetime. The first edition was published as *Anatomy Descriptive and Surgical*, and the subsequent edition as *Anatomy of the Human Body*.

Very little is known about Henry Gray, mostly because he died young, at the age of 34, after contracting smallpox from a nephew for whom he was caring; the nephew survived. Gray was born in London, and seemingly never left. He enrolled as a medical student at St. George's Hospital at the age of 18, his father's work as a messenger to King George IV and King William IV having had some influence. By all accounts, Henry was a diligent, methodical, careful worker who soon realized that his talent was for anatomy, which he learned by doing hands-on dissection for himself.

After qualifying, Gray stayed at St. George's as a demonstrator, then lecturer of anatomy

and also became curator of the hospital's museum. These posts enabled him to continue his anatomical research, and he also began to explore embryology, but we will never know what he might have discovered in that field.

The *Anatomy* was probably conceived as an aid for his students, and although its style is clear and concise, it would be fair to say that without the detailed illustrations that accompany the text, the book would not be so compelling or useful. Following in the tradition of Eustachius and Vesalius, Gray worked closely with an illustrator—his friend, colleague, collaborator, and fellow anatomist, Dr. Henry Vandyke Carter; Carter made the meticulously crafted drawings that were used for the 363 engravings that illustrated the first 750-page edition. The work was so popular that a second edition was published only two years later.

The following year, Gray applied for the post of Assistant Surgeon at St. George's. He would possibly have been appointed but died before the appointment could be made. At the time of his death, he was already at work on the second edition of the book that would make his name immortal.

THE AUTONOMIC NERVOUS SYSTEM

the 30-second anatomy

3-SECOND INCISION
The two branches of the autonomic nervous system —the parasympathetic and the sympathetic—work together to maintain balance in many of the body's processes.

3-MINUTE DISSECTION
The fight-or-flight response is a survival mechanism mediated by the sympathetic pathway. A person is readied to either fight or flee from a stressful or dangerous situation. The heart and lungs are stimulated to work harder and blood vessels open more to parts of the body prepared for action, such as the muscles, while conversely closing more to nonessential areas, such as the bowel.

The autonomic nervous system is separated into parasympathetic and sympathetic branches. The parasympathetic branch is more important in day-to-day activities, while the sympathetic branch is active in stressful situations. Together they help to maintain balance between diverse body processes. Many of the parasympathetic pathways begin in the brain stem and cranial nerves, such as the vagus nerve, while others begin low down in the spinal cord. The nerves travel to near their target tissues, where they communicate with a second nerve in a structure known as a ganglion. From there, the second nerve runs to its target organ, where it may fire off a signal to influence functioning. Parasympathetic functions include slowing the heart rate, increasing digestive and pancreatic secretions, and narrowing the pupil in the eye. The sympathetic pathway begins in the spinal cord. Nerves leaving the spinal cord communicate with a second nerve in a chain of ganglia that lie on each side of the spine. The second nerve travels to its target organ, where effects, such as an increase in the rate and force of contraction of the heart, widening of the pupil, constriction of small arteries, and widening of lung airways, can occur. Nerves may interconnect in structures known as plexuses before then traveling to their organ.

RELATED TOPICS
See also
THE HEART
page 64
THE MAJOR ARTERIES & VEINS
page 66

3-SECOND BIOGRAPHY
WILDER G. PENFIELD
1891–1976
A Canadian neurosurgeon who mapped the sensory and motor cortices of the brain, and invented the Montreal Procedure to treat patients with severe epilepsy

30-SECOND TEXT
Andrew T. Chaytor

The autonomic nervous system regulates body functions—such as the heartbeat and digestive movements— that are not under conscious control.

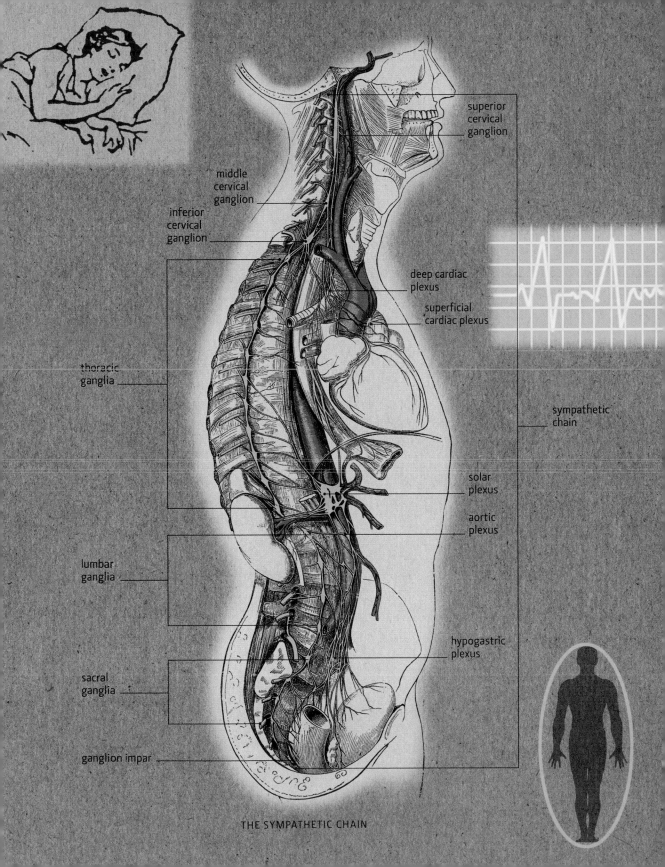

superior
cervical
ganglion

middle
cervical
ganglion

inferior
cervical
ganglion

deep cardiac
plexus

superficial
cardiac plexus

thoracic
ganglia

sympathetic
chain

solar
plexus

aortic
plexus

lumbar
ganglia

hypogastric
plexus

sacral
ganglia

ganglion impar

THE SYMPATHETIC CHAIN

THE CRANIAL NERVES

the 30-second anatomy

Cranial nerves arise directly from the brain, in contrast to spinal nerves, such as the cervical, thoracic, and lumbosacral nerves, which arise from the spine. There are 12 cranial nerves, each identified by a Roman numeral: (I) olfactory, governing the sense of smell; (II) optic, carrying impulses from the retina in the eye to the brain; (III) oculomotor and (IV) trochlear, controlling the eye movements; (V) trigeminal, governing feeling in the face and muscular movements, such as chewing and swallowing; (VI) abducens, another nerve controlling eye movement; (VII) facial, controlling muscles involved in facial expression and conveying tastes; (VIII) vestibulocochlear, which carries balance and sound input from the inner ear to the brain; (IX) glossopharyngeal, supplying nerves to the middle ear, tonsils, and pharynx, among other tasks; (X) vagus, conveying motor and sensory fibers to various places in the head and chest; (XI) accessory spinal, controlling neck and shoulder muscles; and (XII) the hypoglossal nerve, controlling tongue movements. Except for the olfactory nerve, which has its origin in the nasal cavity, all the other cranial nerves originate from nuclei deep within the brain stem (or related structures). They leave the skull cavity through openings called foramina within the bones of the skull.

RELATED TOPICS
See also
THE SKULL
page 22
THE BRAIN & BRAIN STEM
page 128
THE SPINAL CORD
page 130
THE AUTONOMIC NERVOUS SYSTEM
page 134

3-SECOND BIOGRAPHY
GALEN OF PERGAMON
129–ca. 216 CE
Greek-born Roman physician who described 7 out of the 12 cranial nerves

30-SECOND TEXT
December S. K. Ikah

The trigeminal nerve carries sensory and motor fibers; it controls both feeling in the face and muscular actions, such as biting and swallowing.

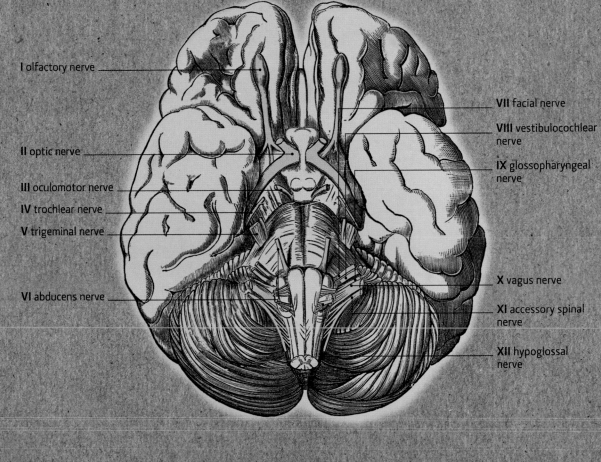

I olfactory nerve

II optic nerve

III oculomotor nerve

IV trochlear nerve

V trigeminal nerve

VI abducens nerve

VII facial nerve

VIII vestibulocochlear nerve

IX glossopharyngeal nerve

X vagus nerve

XI accessory spinal nerve

XII hypoglossal nerve

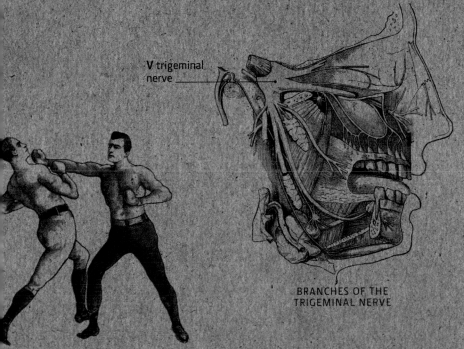

V trigeminal nerve

BRANCHES OF THE
TRIGEMINAL NERVE

THE NERVE PLEXUSES

the 30-second anatomy

3-SECOND INCISION
A nerve plexus is a collection of nerves with many branches emerging from it, and these branches supply muscles or organs surrounding the plexus.

3-MINUTE DISSECTION
When a woman in labor has an epidural, anesthetists inject anesthetic agents into her lower back, around the lumbar spine. The agent prevents pain by blocking nerves from the lumbar and sacral plexuses. Because nerve impulses from these plexuses are blocked, the woman's pelvis and lower limbs are numbed. Minor operations, such as hand surgery, can be conducted by blocking the brachial plexus at the shoulder instead of putting a patient to sleep.

A nerve plexus has a similar construction to a subway. Subway networks have multiple train lines that merge and disperse at various stations, in various suburbs, and along various routes; any one station may have several lines running through it. Within a nerve plexus, individual nerves emerge from networks to supply particular structures. The networks are formed by small spinal nerves, nerves that exit directly from the spinal cord. These spinal nerves have multiple fibers that can contribute to different individual nerves emerging from the plexus. Each plexus corresponds to a different body system. For example, the spinal nerves that emerge directly from the spinal cord within the neck are called cervical nerves. These cervical nerves form the brachial plexus, which supplies the upper limb. The upper limb could be likened to a particular suburb of a city; different nerves (trains) stop at different muscles (stations) around the arm (suburb). In addition to the brachial plexus, other major nerve plexuses include: the cervical plexus, which supplies the head and neck, and the lumbar and sacral plexuses that supply the lower limb and pelvis.

RELATED TOPICS
See also
THE SPINAL CORD
page 130
THE AUTONOMIC
NERVOUS SYSTEM
page 134

3-SECOND BIOGRAPHY
LEOPOLD AUERBACH
1828–1897
A German anatomist and one of the first to investigate the nervous system using histological staining methods. His name is associated with a layer of ganglia called Auerbach's plexus

30-SECOND TEXT
Gabrielle M. Finn

Cervical nerves from the spinal cord pass through and combine in the brachial plexus on their way to supply many muscles in the upper limb or arm.

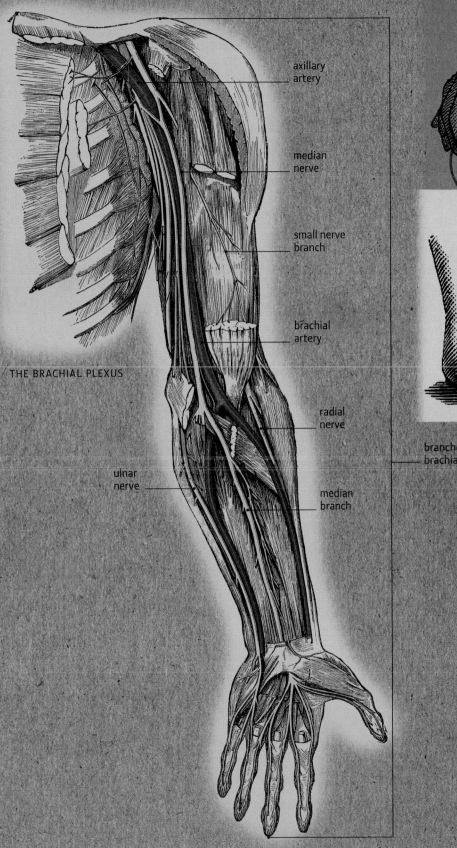

axillary
artery

median
nerve

small nerve
branch

brachial
artery

THE BRACHIAL PLEXUS

radial
nerve

ulnar
nerve

median
branch

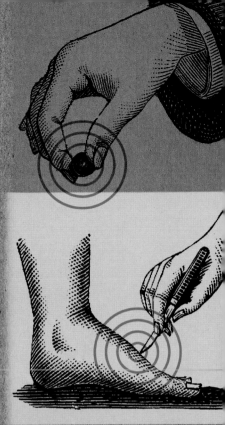

branches of the
brachial plexus

THE REPRODUCTIVE SYSTEM

THE REPRODUCTIVE SYSTEM
GLOSSARY

cervix Lower area of the uterus that narrows to join the vagina. The cervix takes its name from the Latin *cervix uteri*, meaning "neck of the womb."

glans clitoris An organ formed of sensitive erectile tissue situated where the labia joins below the pubic bone.

glans penis Bulbous sensitive end of the penis; at its tip is the opening of the urethra (urinary meatus) through which urine and semen exit the body. When the penis is not erect, the glans is covered by the foreskin in males who have not been circumcized. In females, the anatomical equivalent of the glans penis is the glans clitoris.

labia Folds of skin covering the entrance to the vagina; labia majora are outer folds and labia minora inner folds.

levator ani With the coccygeus muscle, one of the pelvic floor muscles in the pelvic diaphragm; it supports organs, such as the uterus, ovaries, and fallopian tubes, and makes sure that the vagina and rectum are closed. The pelvic floor muscles play a key role in maintaining continence by acting as a sphincter for the urethra.

ovary Principal organ of the female reproductive system, in which ova (egg cells) and female sex hormones, including estrogen and progesterone, are produced. A small number of ova will reach maturity and are released from a Graafian follicle in the ovary into the fallopian tubes, where they may be fertilized by sperm. A woman has two ovaries, one on each side of the uterus in the abdomen.

ovum An egg released by one of a woman's ovaries in the process of ovulation. An immature ovum is called an oocyte. If fertilized in a fallopian tube, the egg becomes a zygote, then as it divides it becomes known as a morula. When it embeds in the wall of the woman's uterus it is called a blastocyst. After it has embedded, the blastocyst is known as an embryo.

prostate gland Accessory male sex gland that secretes an alkaline fluid during ejaculation that is, with sperm, a key constituent of semen. The prostate gland empties into the urethra where it leaves the bladder. Ejaculate passes down the urethra to the tip of the penis and out of the body. A typical ejaculation contains up to 300 million sperm, but in volume terms, the sperm account for only 5 percent of the semen.

testis Male sex gland, also known as testicle, in which sperm and the male hormone testosterone are produced. Typically males have a pair of testes; they develop within the abdomen of a fetus, but before birth descend into a skin pouch (scrotum) behind the penis. Sperm produced in the testis pass along the vasa efferentia to be stored and to mature in the epididymis. After puberty, the testes produce around 1,000 sperm every minute, each $1/500$ inch (0.05 mm) in length and taking ten weeks to become fully mature.

urethra Tube leading from the bladder to the body's exterior, passing down inside the penis in males and to the front of the vagina in females. Around 8 inches (20 cm) long in men and $1\frac{1}{2}$ inches (4 cm) long in women, it carries urine in both sexes and in men also semen.

uterus Also known as the womb, the part of the female reproductive system in which a fertilized egg is embedded, and in which an embryo and fetus may develop. About five days after fertilization the egg (now known as a blastocyst) embeds in the wall of the uterus; its outer cells form the placenta and other cells form the embryo. The uterus is in the center of the pelvis, between the bladder to the front and the rectum at the rear.

vagina The lowest part of the female reproductive tract, a muscular canal lined with mucous membrane that runs from the cervix to the vestibule of the external female genitalia. During sexual intercourse, the erect penis ejaculates semen into the upper part of the vagina, from where sperm swim through the cervix to fertilize the egg in a fallopian tube. In childbirth, the baby passes through the cervix and vagina.

THE FEMALE REPRODUCTIVE SYSTEM

the 30-second anatomy

3-SECOND INCISION
The female reproductive system has several functions—to produce eggs, secrete hormones, receive sperm, provide a site for implantation, then protect and deliver a baby.

3-MINUTE DISSECTION
At ovulation, an oocyte (immature ovum) is shed from the surface of an ovary and, normally, immediately enters the open end of a fallopian tube. A funnel-shaped part of the tube (infundibulum) opens near the ovary, and elongated processes (fimbriae) from the tube extend over the ovary, with the longest attached to the surface of the ovary. Fimbriae appear to guide the ovum into the inner part (lumen) of the fallopian tube.

The primary female reproductive organs are the ovaries and accessory ducts—the fallopian tube, uterus, and vagina. The ovaries produce eggs and female sex hormones, which—among other things—promote the development of breasts and regulate a woman's menstrual cycle. Each ovary is adjacent to the open end of a fallopian tube; when released from the ovary, an egg passes down the tube. The fallopian tubes are narrow and about 4 inches (10 cm) long: they open into the uterus. The uterus is a thick-walled muscular organ. In a woman who has never been pregnant, the uterus is approximately 3 inches (7.5 cm) long, $1/2-3/4$ inch (1–2 cm) thick, and about 2 inches (5 cm) at its widest, and lies entirely within the pelvic cavity. During pregnancy, the uterus becomes greatly enlarged and extends into the abdomen. That part of the uterus above the openings of the fallopian tubes is termed the fundus. Below the fundus is the uterus body, which, lower down, becomes the cervix of the uterus. The vagina is a thin-walled tube extending from the cervix and opening into the vestibule of the external genitalia.

RELATED TOPICS
See also
THE PELVIS
page 28

THE MALE
REPRODUCTIVE SYSTEM
page 150

3-SECOND BIOGRAPHY
REGNIER DE GRAAF
1641–1673
A Dutch physician who was the first to describe Graafian follicles, ripe or mature follicles on the ovary that are preparing to release an egg

30-SECOND TEXT
Claire France Smith

During pregnancy, the walls of the uterus greatly expand to accommodate the growing baby, and their strong muscle layer becomes thinner.

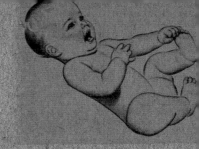

THE FEMALE REPRODUCTIVE ORGANS

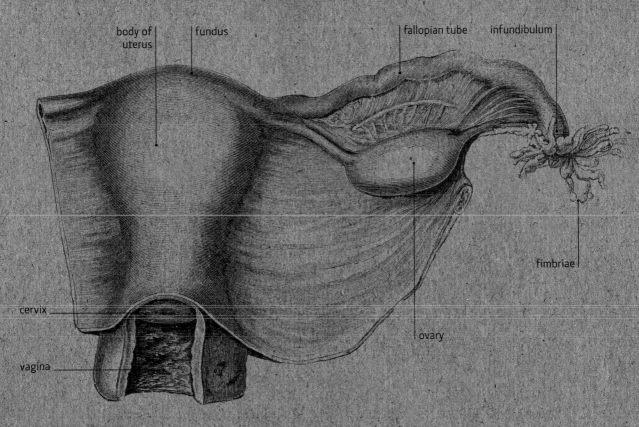

body of uterus

fundus

fallopian tube

infundibulum

fimbriae

cervix

vagina

ovary

OVUM (EGG)

CROSS SECTION SHOWING THE UTERUS IN SITU

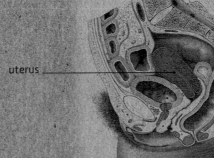

uterus

THE PELVIC FLOOR MUSCLES

the 30-second anatomy

3-SECOND INCISION
The pelvic floor is a wall between the pelvis and perineum that functions to support the pelvic organs and has a role in maintaining continence.

3-MINUTE DISSECTION
Muscles of the pelvic floor can tear and become damaged during childbirth. When a baby is being delivered, a physician may choose to cut through the muscles of the pelvic floor and the skin between the vagina and anus to prevent tearing—this process is known as an episiotomy. If the perineal body tears, the woman can have a prolapse of the vagina, rectum, or bladder.

The pelvic floor is formed by the perineal membrane and muscles of the pelvic diaphragm and deep perineal pouch. Of these, the pelvic diaphragm is the dominant muscular part. It consists of the coccygeus muscle and the levator ani, formed by the muscle fibers of pubococcygeus, puborectalis, and iliococcygeus. The fibers of the levator ani run from the bony walls of the pelvis toward the midline, where they merge. In both sexes, this union occurs behind the anal opening, and in females it is also behind the vagina. At the front of the pelvis the muscles are separated by a gap (urogenital hiatus) that lets the urethra exit the pelvis; in females, the vagina also passes through this gap. The levator ani supports the pelvic organs (the uterus, ovaries, and fallopian tubes) and keeps the vagina and rectum closed. The muscles beneath the diaphragm (in the deep perineal pouch) act as a sphincter for the urethra and stabilize the perineal body—a connective tissue structure onto which the pelvic floor muscles attach. The perineal body maintains the integrity of the pelvic floor. Pelvic floor muscles can be strengthened by Kegel exercises, which are particularly useful for maintaining urinary continence.

RELATED TOPICS
See also
THE BLADDER
page 98
THE FEMALE
REPRODUCTIVE SYSTEM
page 144
THE MALE
REPRODUCTIVE SYSTEM
page 150
THE PERINEUM
page 152

3-SECOND BIOGRAPHY
ARNOLD KEGEL
1894–1981
An American gynecologist who invented Kegel exercises to strengthen the pelvic floor

30-SECOND TEXT
Gabrielle M. Finn

The two parts of the levator ani combine to the rear of the anal opening and the tissue of the perineal body gives strength to the pelvic floor muscles.

urethral opening

vaginal opening

transverse
perineal muscles

perineal body

levator ani

gluteus maximus

FEMALE PELVIC FLOOR

sacrum

piriformis

coccygeus

levator ani

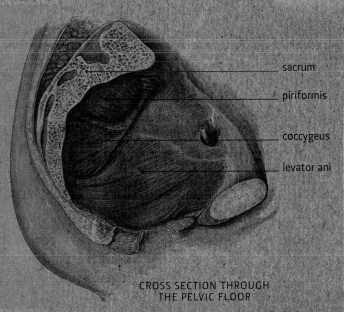

CROSS SECTION THROUGH
THE PELVIC FLOOR

1718
Born at Long Calderwood, South Lanarkshire, Scotland

1762
Became Physician Extraordinary to Queen Charlotte

1770
Built a house in Glasgow, now forms part of the Hunterian Museum and Art Gallery

1775
Commissioned sculpture of a flayed corpse as a teaching aid

1731
Graduated in Divinity from the University of Glasgow

1767
Elected Fellow of the Royal Society

1774
Published *Anatomia uteri umani gravidi* ("Anatomy of the Gravid Human Uterus") with engravings inspired by the drawings of Leonardo da Vinci

1783
Died in London

1737
Began medical studies at Edinburgh

1768
Elected Professor of Anatomy to the Society

1741–44
Became a pupil of William Smellie, specializing in obstetrics, at St. George's Hospital, London

1768
Opened his own school, built the anatomy theater and museum in Great Windmill Street, London

1743
Published the paper "On the structure and diseases of articulating cartilages"

1769–72
Professor of Anatomy at the Royal Academy of Arts

1746
Began teaching private anatomy and surgery classes

1756
Became a licensed physician

WILLIAM HUNTER

William Hunter, anatomist, physician, and obstetrician, united a love of medicine with a passion for the arts. He was not only a Fellow of the Royal Society, but also Professor of Anatomy at the Royal Academy of Art. When he published his masterwork *Anatomy of the Gravid Human Uterus* in 1774, he chose the drawings of Leonardo da Vinci as exemplars of the clear illustration style he wanted; he had access to Leonardo's work because the drawings were in the royal collection at Windsor Castle and he was by then Physician Extraordinary to Queen Charlotte (mother of 15 children, 13 of whom survived into adulthood—perhaps a testament to Hunter's obstetric skill). So it is fitting that his monument is the Hunterian Museum and Art Gallery (now part of Glasgow University), based on the house he built in 1770.

Possessed of a discriminating intelligence as well as charm, refinement, and a ferocious work ethic, Hunter began his adult life as a student of divinity, then studied medicine at Edinburgh before moving to St. George's Hospital in London in 1741. Here, he became pupil to his fellow Scot William Smellie, who specialized in obstetrics and pioneered the use of forceps; Hunter eschewed them. After flirting with orthopedics, Hunter quickly became the foremost obstetrician in London, with a very distinguished clientele. However, anatomy was his first and enduring love. He set up his own school of anatomy and surgery, assisted by his younger brother John, to whom he was mentor and teacher. William introduced the practice (normal in France) of providing each student with their own cadaver to dissect, and commissioned Italian sculptor Agostini Carlino to make a cast of a flayed corpse as an accurate yet aesthetically pleasing teaching aid.

Hunter's medical practice and his teaching brought success, and great wealth. He was an avid collector of books and antiquities, and bequeathed his collections to the nation—they are now housed in the Hunterian Museum and Art Gallery. Unfortunately, his brother John, whose even more glittering career in pathological anatomy William had helped ignite, was possessed of less emollient character, and quarreled with his brother three years before William's death. The rift was never healed.

THE MALE REPRODUCTIVE SYSTEM

the 30-second anatomy

The male reproductive system
consists of a network of tubes, beginning from
the testis, and three glands that have grown
out of the tubes during development. Its main
purpose is to produce and store fertile sperm
and then deliver these sperm through the tubes
into the female reproductive system; in addition,
the development of physical male characteristics,
such as facial hair, is dependent on hormones
secreted by cells in the testis. Within each
testis, tiny tubes are lined by two types of
cells—the germ cells that produce sperm and
the sertoli cells that support developing sperm.
Outside the tubes in the testis lie Leydig cells;
these produce testosterone, which regulates
sperm production. Sperm are moved into the
epididymis for storage. From the epididymis, the
sperm are transported via the ductus deferens
and the ejaculatory duct into the urethra that
runs to the tip of the penis. When stimulated,
the secretions of the bulbourethral gland
lubricate the urethral tube, while the seminal
and prostate glands bathe the sperm in seminal
fluid. The penis, which delivers semen into the
vagina, is surrounded by three spongy cylinders
that fill up with blood to make the penis erect.

*Within the scrotum,
the epididymis is a
coiled tube in which
newly produced semen
mature and are stored;
it connects the testis
to the ductus deferens.*

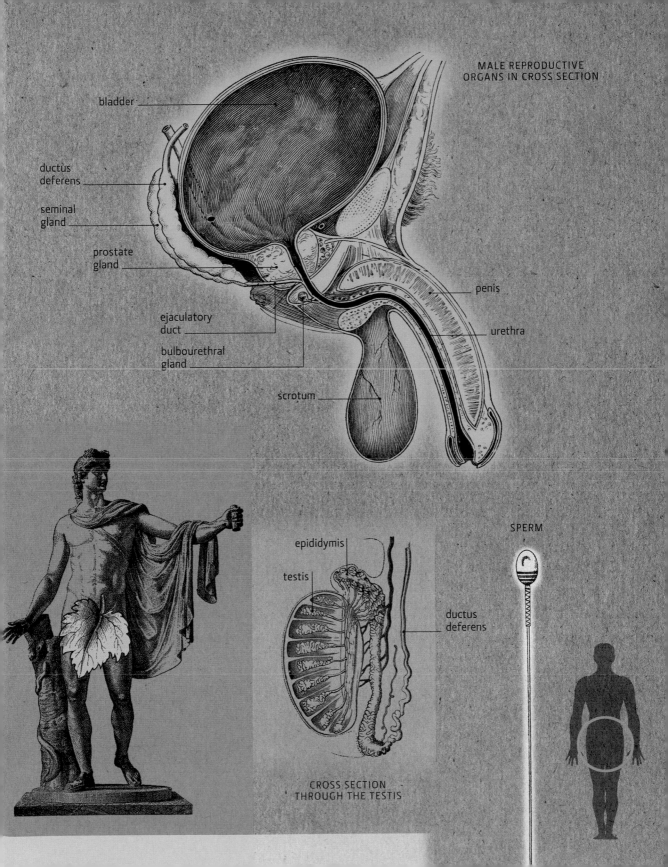

MALE REPRODUCTIVE
ORGANS IN CROSS SECTION

bladder

ductus
deferens

seminal
gland

prostate
gland

ejaculatory
duct

bulbourethral
gland

scrotum

penis

urethra

epididymis

testis

ductus
deferens

SPERM

CROSS SECTION
THROUGH THE TESTIS

THE PERINEUM
the 30-second anatomy

3-SECOND INCISION
The perineum is the diamond-shaped area in which the male and female genital organs, urinary tract, and anus open.

3-MINUTE DISSECTION
Animal studies show that the anogenital distance (AGD)—the distance between the anus and the rear base of the scrotum—correlates with fertility in male rodents; the shorter the AGD, the lower the fertility. In addition, shortening of AGD in rodents is related to exposure to toxic chemicals during pregnancy. A 2011 human study showed that men with a very short AGD have lower sperm count, poor sperm quality, and lower concentration of sperm.

The bones on which a person sits are the ischial bones, situated on both sides of the hip. The pubic bone in front and the tailbone (coccyx) behind are both raised above the ischial bones. Together, these bones form the diamond-shaped outline of the perineum. A line between the ischial bones divides the perineum into two triangles. The urogenital triangle in front contains the penis and scrotum in males, and the vagina in females; behind this line in both male and female is the anal triangle containing the anus and its opening. The roof of the perineum, a sheet of two muscles, is the pelvic diaphragm, while the floor consists of skin and fibrous tissue. In the urogenital triangle, the roof is reinforced by fibrous tissue called perineal membrane. The fibrous tissue of the roof and floor are continuous and together create the outer space of the perineum. Within this space lie the roots of the penis or clitoris, glands, and muscles. The muscles of the pelvic diaphragm, the fibrous perineal membrane, and deep fibrous tissues are all attached to a tough tissue called the perineal body, which lies just in front of the anal opening.

RELATED TOPICS
See also
THE FEMALE
REPRODUCTIVE SYSTEM
page 144
THE PELVIC FLOOR MUSCLES
page 146
THE MALE
REPRODUCTIVE SYSTEM
page 150

3-SECOND BIOGRAPHY
SIR RUTHERFORD ALCOCK
1809–1897
A British surgeon who described the canal through which the pudendal nerve, artery, and vein pass in and out of the perineum

30-SECOND TEXT
December S. K. Ikah

The diamond-shaped perineum that links the pubic bone (front), coccyx (rear), and two ischial bones (side) is typically twice as long in men as in women.

MALE PERINEUM

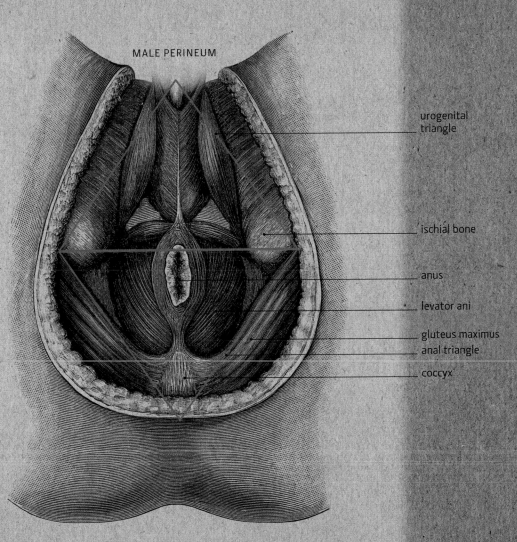

urogenital
triangle

ischial bone

anus

levator ani

gluteus maximus

anal triangle

coccyx

THE PERINEAL SPACE

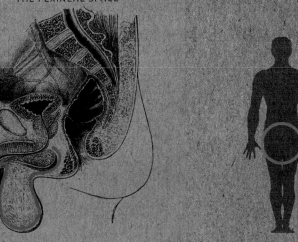

fibrous layer of
the perineal floor

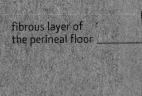

APPENDICES

RESOURCES

BOOKS

Atlas of Human Anatomy
Frank H. Netter
(Saunders; 5th edn, 2010)

Clinically Oriented Anatomy
Keith L. Moore, Arthur F. Dalley,
and Ann M. R. Agur
(Lippincott Williams & Wilkins;
6th edn, 2009)

Grant's Dissector
Patrick W. Tank
(Lippincott Williams & Wilkins; 15th edn,
2012)

Gray's Anatomy
Henry Gray
(1858; 40th edn, Arcturus Publishing, 2010)

Gray's Anatomy for Students
Richard L. Drake, A. Wayne Vogel,
and Adam W. M. Mitchell
(Churchill Livingston, 2010)

Human Physiology: An Integrated Approach
Dee Unglaub Silverthorn
(Benjamin Cummings, 2009)

Neuroanatomy: An Illustrated Color Text
Alan R. Crossman and David Neary
(Churchill Livingston, 2010)

WEB SITES

The American Association of Anatomists
http://aaatoday.org/
The AAA is an organization dedicated to
the advancement of anatomical science
through research, education, and professional
development activities.

The Anatomical Society
http://www.anatsoc.org.uk/
The Anatomical Society, founded in 1887, is a
learned society with charitable status. Its aims
are to promote, develop and advance research
and education in all aspects of anatomical
science. It achieves these aims by organizing
scientific meetings; publishing the *Journal of
Anatomy and Aging Cell*; making annual awards
of PhD studentships, grants, and prizes.

Instant Anatomy
http://www.instantanatomy.net/
Instant anatomy is a specialized web site
devoted to human anatomy with diagrams,
podcasts, and revision questions.

Visible Body
http://www.visiblebody.com/
A comprehensive visualization tool, *Visible Body*
is a virtual human anatomy web site containing
highly detailed, anatomically accurate, 3D models
of all the human body systems. Also includes
tutorial videos.

NOTES ON CONTRIBUTORS

Judith Barbaro-Brown is Teaching Fellow at the School for Medicine & Health at Durham University. Originally a podiatrist, she worked within the National Health Service before moving into teaching podiatric medicine. She now teaches muscular skeletal anatomy, clinical skills, and histology, and acts as Senior Educational Advisor to a number of professional bodies in the UK.

Jo Bishop is Curriculum Director for the Graduate Entry in Medicine (GEM) program at the College of Medicine, Swansea University. She oversees the GEM curriculum's design, structure, and development to ensure educational alignment and adherence to contemporary medical education principles. Jo Bishop holds a PhD and also lectures in anatomy.

Andrew Chaytor has a PhD in Cardiovascular Physiology and Pharmacology from the University of Wales, Cardiff. He is a Lecturer at the School of Medicine and Health at Durham University, where he also conducts research into medical education. Previously he taught anatomy, physiology, and pharmacology at the Sunderland School of Pharmacy.

Gabrielle M. Finn is a Lecturer in Anatomy at Durham University. She has a PhD in Medical Education. Gabrielle researches anatomy, pedagogy, and medical professionalism, and teaches anatomy and clinical skills. She is a member of council for the Anatomical Society, a member of the Federative International Programme for Anatomical Education (FIPAE), and a fellow of the Center for Excellence in Teaching and Learning (CETL).

December S. K. Ikah practiced general medicine along with a Lectureship in Human Anatomy at Niger Delta University, Nigeria. He holds a PhD in Neurotoxicity of nanomaterials from Liverpool University. As Teaching Fellow at the School of Medicine and Health, Durham University, Dr. Ikah teaches human anatomy and clinical skills, and conducts research into medical education.

Marina Sawdon is a cardiovascular physiologist with a PhD from Durham University, where she is a Physiology Lecturer in cardiovascular, respiratory, and renal medicine. She was co-physiology editor for the *Anaesthesia and Intensive Care Medicine Journal*. Her areas of research have included the field of whole body cardiovascular physiology and, more recently, medical education.

Claire France Smith is a member of the Faculty of Medicine at Southampton University. She is a Senior Teaching Fellow in the Center for Learning Anatomical Sciences. She has taught anatomy to medical, dental, and allied health professions from entry level to examining for the Royal College of Surgeons. She is a member of council for the Anatomical Society and Chair of the Education Committee.

INDEX

ACKNOWLEDGMENTS

PICTURE CREDITS
The majority of the illustrations in this book were produced using images from *Anatomy of the Human Body* by Henry Gray, F.R.S. (1918) and *Anatomy Descriptive and Surgical* by Henry Gray F.R.S. (1905).

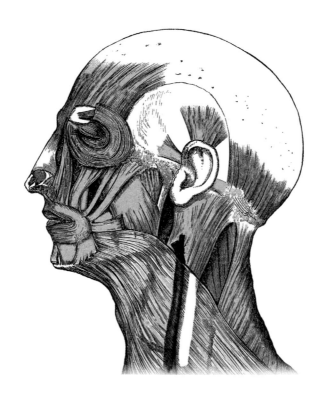